Introduction to Experimental Biophysics

A LABORATORY GUIDE

Introduction to Experimental Biophysics

A LABORATORY GUIDE

Jay L. Nadeau

CRC Press
Taylor & Francis Group
Boca Raton London New York

CRC Press is an imprint of the
Taylor & Francis Group, an **informa** business

CRC Press
Taylor & Francis Group
6000 Broken Sound Parkway NW, Suite 300
Boca Raton, FL 33487-2742

First issued in hardback 2021

Version Date: 20141117

ISBN 13: 978-1-138-40738-1 (hbk)
ISBN 13: 978-1-4665-5765-9 (pbk)

Publisher's Note
The publisher has gone to great lengths to ensure the quality of this reprint but points out that some imperfections in the original copies may be apparent.

Library of Congress Cataloging-in-Publication Data

Nadeau, Jay L., author.
[Introduction to experimental biophysics (2012)]
Introduction to experimental biophysics : a laboratory guide / Jay L. Nadeau.
p. ; cm.
Complemented by: Introduction to experimental biophysics : biological methods for physical scientists / Jay L. Nadeau. c2012.
Includes bibliographical references and index.
ISBN 978-1-4665-5765-9 (alk. paper)
I. Nadeau, Jay L. Introduction to experimental biophysics. Complemented by (work): II. Title.
[DNLM: 1. Biophysics--methods--Laboratory Manuals. 2. Biochemistry--methods--Laboratory Manuals. 3. Molecular Biology--methods--Laboratory Manuals. QT 25]

QH505
612'.014--dc23
2014044955

Visit the Taylor & Francis Web site at
http://www.taylorandfrancis.com

and the CRC Press Web site at
http://www.crcpress.com

Contents

Section Two: Gene Expression

Section Three: Fluorescent Reporter Genes in Mammalian Cells

Section Four: Nanoparticles

Section Five: Electrophysiology

Appendices: Instructor Information and Guides

Acknowledgment

Thanks to all of the students who took Biomedical Engineering 506 at McGill and who helped to improve the course with their ideas and suggestions. Also, thanks to my editor, Luna Han, who encouraged the idea of this book and made it a reality.

About the Author

Jay L. Nadeau is an associate professor of biomedical engineering and physics at McGill University (2004–present) whose research interests include nanoparticles, fluorescence imaging, and development of instrumentation for detection of life elsewhere in the solar system. Her group was the first to label bacteria with quantum dots and to explore the possibility of using fluorescent labels as tools for detection of traces of extraterrestrial life. Every year, she travels to the Canadian High Arctic to do fieldwork at the McGill Arctic Research Station (MARS), a Mars analog site at nearly 80° north latitude. She has published over 50 papers on topics ranging from theoretical condensed matter physics to experimental neurobiology to development of anticancer drugs, in the process using almost every single one of the techniques described in this book. Her work has been featured in *New Scientist, Highlights in Chemical Biology*, Radio Canada's *Les Années lumière, Le Guide des Tendances*, and in educational displays in schools and museums. Her research group features chemists, microbiologists, roboticists, physicists, and physician-scientists, all learning from each other and hoping to speak each other's language. A believer in bringing biology to the physicists as well as physics to the biologists, she has created two graduate-level courses: Methods in Molecular Biology for Physical Scientists and Mathematical Cellular Physiology. She also teaches pharmacology in the medical school and was one of the pioneers in the establishment of the multiple mini-interviews (MMIs) for medical school admission. She has an adjunct position with the Jackson Laboratory in Bar Harbor, Maine, and collaborators in industry and academia in the United States, Europe, Australia, and Japan. She has given several dozen invited talks at meetings of the American Chemical Society, American Geophysical Union, SPIE (International Society for Optics and Photonics), the Committee on Space Research (COSPAR), and many others. Before McGill, she was a member of the Jet Propulsion Laboratory's Center for Life Detection; previous to that, she was a Burroughs-Wellcome postdoctoral scholar in the laboratory of Henry A. Lester at Caltech. She received her PhD in physics from the University of Minnesota in 1996.

Preface: How to Use This Manual

This book is intended as a laboratory companion to *An Introduction to Experimental Biophysics: Biological Methods for Physical Scientists* (Boca Raton, FL: Taylor & Francis 2012). It is structured according to the chapters in the textbook as much as is feasible. A series of 14 chapters presents a wide variety of techniques that may be performed during a semester-long, 3-credit course or during a 1-month intensive. The manual is organized to follow a logical experimental progression beginning with a DNA construct. The construct is amplified, mutated, or fused to another gene; expressed in bacteria; and purified as either a protein or plasmid DNA. The purified protein is characterized using spectroscopy, crystallization, and lipid bilayer expression. The purified plasmid DNA is expressed in mammalian cells and characterized by advanced imaging techniques and electrophysiology. The laboratories also include two sessions on nanoparticle characterization and synthesis.

All of the experiments in the manual have been tested in a pedagogical setting and found to be appropriate in terms of time, cost, and complexity.

The book is aimed primarily at instructors of laboratory courses in biophysics/molecular biology at the advanced undergraduate or graduate level. This manual will allow instructors to schedule a full-term or summer course based on the textbook. The manual is intended to aid course design and planning by suggesting sample experiments that use a select set of commercially available reagents; by scheduling the experiments so that gaps occur during weekends or lecture/computer lab time; and by providing a guide to results and pitfalls for each experiment.

Expected results and pitfalls are given for each experiment. End-of-chapter questions and problems are added to reinforce conceptual understanding. The DNA construct chosen as an example is the *Aequorea victoria* green fluorescent protein (GFP) gene and several of its variants. However, the general structure of the laboratories can be readily adapted to other constructs and proteins besides GFP. A section on advanced options is included to help guide instructors who wish to go beyond the simple example constructs. The course may be as simple or as sophisticated as time, students, and infrastructure warrant. One or more of the experiments may also be performed by students on their own if they wish to learn techniques without a formal course. Although all of the molecular biology experiments must be done in sequence, the more advanced experiments (transfection, nanoparticle synthesis, electrophysiology) stand alone.

The chapters are organized according to a schedule that permits completion of the laboratories in a 13-week quarter or 3-week intensive summer term, with minimal dead time. This schedule has been refined over 3 years of teaching the course and has been found to work well.

The experiments are designed to be performed by teams of two students; each team performs all of the experiments together throughout the term (notebook entries and reports should be written by individuals). Some experiments are divided into parts A and B for scheduling reasons. An experiment designated Part B must be done after a certain interval of time to allow something to grow or develop but is usually fast (<30 minutes to complete). In a summer course, A and B sections may be done on subsequent days. In a course given during the year when there is only one laboratory session per week, the B parts must be specially scheduled or performed by teaching assistants.

Introduction: Infrastructure and Safety Requirements

I.1 Infrastructure

A basic teaching wet lab containing bench space, sinks, electrical outlets, and appropriate emergency facilities (safety shower, eyewash) is needed for all of the experiments in this book.

The molecular cloning and mammalian cell experiments require access to autoclaves for the sterilization of glassware and waste. A refrigerator, freezer, and cryogenic (–80°C) freezer are needed for storing plasmids and reagents.

The experiments involving mammalian cell lines (Experiments 9 and 10) require access to a laminar flow hood (safety cabinet) and CO_2 incubators. A liquid nitrogen Dewar is needed if mammalian cell lines are to be preserved. Imaging of transfected mammalian cells requires fluorescence microscopy. If a laboratory containing teaching instruments is available, this is ideal; cells may be fixed and transported if necessary. Alternatively, for class sizes that are small enough, a single demonstration instrument can be used and students can take turns. I have access to only one instrument and have taught class sizes of up to 20 students. In this case, it is desirable to split the class—for example, half the class can do Experiments 10 and 11 in one week and the other half Experiments 12 and 13. The following week, the participants perform the reverse set of experiments.

Experiment 14 (on quantum dots) is best performed as a demonstration in a laboratory equipped for anoxic synthesis. The students focus on characterization of the nanoparticles after synthesis, which can include spectroscopy, determination of quantum yield, or fluorescence microscopy.

Experiment 16 (on electrophysiology) requires a bilayer chamber and electrophysiology amplifier. This experiment should only be performed if these instruments are available in the instructor's laboratory or a specialized teaching laboratory. I perform this experiment as a demo in the same way as with the mammalian cells.

A few pieces of equipment are necessary for nearly all experiments. These are usually available in teaching laboratories:

- Shaking incubator for *Escherichia coli* (one large incubator per class)
- Bacterial incubator or warm room at 37°C (one per class)
- Absorbance spectrometer or plate reader (one per class)
- Fluorescence plate reader or spectrometer (one per class)

- Water baths (37°C, 42°C, and 60°C) or hot plates (one each per class)
- Electroporator if electrically competent cells are used
- Microcentrifuges (one for two teams)
- Vortex (one for two teams)
- Agarose gel box with power supply (one per team)
- Microwave oven for melting agarose (one or two per class)
- Sodium dodecyl sulfate polyacrylamide gel electrophoresis (SDS-PAGE) gel box and casting tray (one per team)
- Light box and software for gel imaging

Some of the exercises require computer access. As most students have their own laptops, these "labs" may be performed in a lecture hall with students using their own computers. In this case, freeware can be used; links are provided in the appropriate chapters to free molecular biology software for the Mac and personal computer (PC). An Internet connection helps for use of online tools. It may also be desirable to have at least one demo computer containing commercial software if such software is used by the instructors. This is a matter of personal preference discussed further in Chapter 3.

I.2 Safety Concerns and Procedures

I.2.1 Hazardous Chemicals

A number of hazardous chemicals are used in the exercises in this book. The most notable are acrylamide (a neurotoxin);

ethidium bromide and other DNA intercalating dyes (believed to be carcinogens); and solvents such as phenol, ethanol, and acetone. For these and other hazardous chemicals, up-to-date material safety data sheets (MSDSs) must be obtained and stored in the lab. An MSDS contains the chemical name, Chemical Abstracts Service (CAS) number, and data on health hazards (including first aid treatment), physical properties, fire and explosion hazards, reactivity, and procedures for a spill or leak. Your institution may also have specific rules about purchasing, storing, and cataloging hazardous chemicals and about training required for personnel handling them. All students should be informed of the location of the MSDSs and of the safety shower and eyewash and instructed on procedures to follow in case of a spill. Lab coats, latex or nitrile gloves, and safety goggles must be worn by all students at all times, along with appropriate attire (long pants and closed-toed shoes). It is recommended that personal items such as backpacks and cell phones be stored in lockers and not taken into the chemistry lab. Solvents should be handled in a chemical fume hood.

Disposal of chemicals and reagents must be done responsibly and students instructed regarding the location of containers for disposal of each hazard. Solvent waste bottles should be located in the fume hood. Waste contaminated with ethidium bromide should be collected in a labeled container. Appropriate containers for pipette tips, sharps, and broken glass should also be provided and labeled.

I.2.2 Physical Hazards

Electrophoresis units are run at high voltage and can cause a serious shock, even electrocution. Students should be warned to always close lids before turning on the power supply and to unplug the power supply before removing a finished gel.

The ultraviolet (UV) light used in gel imaging can cause corneal burns. Safety goggles and a UV-protective face shield should be used if exposure to a transilluminator is necessary. Ideally, the camera should be mounted on a cover so that direct exposure to the lamps does not take place.

To prevent needle sticks, needles should never be recapped, and all used sharps should be placed into a puncture-proof bin.

Bunsen burners carry a risk of fire and must be handled carefully. Alternatively, bacteria may be plated using reusable plastic spreaders, which are collected and autoclaved by instructors.

I.2.3 Biosafety Containment

I.2.3.1 Bacteria

Molecular cloning involves the use of live bacteria. Although cloning strains of *E. coli* are quite fragile and extremely unlikely to pose a health risk, they are disposed of as biohazardous material under biosafety containment level 1 (BSL-1) rules. Containers for biohazards should be placed throughout the lab and students instructed in their proper use. Petri dishes with bacterial colonies, LB (Luria-Bertani) medium with bacterial growth, and similar items must be placed in these containers.

Gloves, tubes used for DNA preparation, and the like are **not** biohazards and do not need to be put into biohazard bins.

I.2.3.2 Mammalian Cells

The cell lines used in these labs are BSL-1. However, some cell lines must be handled under BSL-2 precautions because they are known to be contaminated with oncogenic viruses. This includes the ubiquitous HeLa cell line. It is wise to assume that all mammalian cell cultures may harbor viruses and to treat them accordingly. These precautions will also help protect the cells from contamination.

The exact procedures for handling mammalian cells will need to be approved by the biosafety committee of the institution where they are used. The following are general guidelines for how potentially infectious material should be handled:

- Reduce aerosol formation by using aerosol-free pipettes and tips.
- Place used tips into dilute bleach solution after use.
- Decontaminate work surfaces before and after use (usually with 70% ethanol).
- Autoclave solid waste and disinfect liquid waste (such as used medium) with bleach.
- Most important, use an approved, inspected biological safety cabinet (laminar flow hood) to handle mammalian cells. Such a cabinet has an air barrier in front and a HEPA

(high-efficiency particulate air) filter for exhaust air. Any instructors should be fully trained in the use of the cabinet and students instructed regarding the proper ways to maintain laminar flow while working in the cabinet.

I.2.4 Sterile Technique

It is nearly as important to protect your experiment from you as you from it. Glassware used in molecular biology must be carefully cleaned to remove traces of detergent, dirt, and bacteria that can degrade nucleic acids and inhibit enzymes. All glassware should be thoroughly rinsed, preferably in a dishwashing facility, and autoclaved. For experiments using RNA, all water used must be treated with diethyl pyrocarbonate (DEPC) to inhibit RNases.

Pipette tips or microcentrifuge tubes that are purchased nonsterile should be autoclaved. Pipette tips should be packed into boxes before autoclaving; microcentrifuge tubes may be placed into glass jars or beakers, covered with foil, and autoclaved. They are then removed one by one using tweezers.

For work with mammalian cells, meticulous sterile technique is required. It is recommended that a lecture period be dedicated to the proper handling of cells and the use of laminar flow cabinets before the laboratory is performed. Students should be provided with their own small aliquots of media and other reagents to avoid cross-contamination among teams.

Any media, cells, or reagents showing bacterial or mold growth must be bleached and autoclaved immediately.

I.3 Use and Maintenance of Micropipettes

Biological experiments require accurate transfer of microliters of solutions. In my experience, problems with pipetting represent one of the major barriers to learning molecular biology because they are often not suspected as a source of error and thus go undetected, ruining successive experiments. It is highly recommended that at least a 1-hour period be spent describing the function and use of micropipettes and allowing the students to test their technique using a microbalance.

Micropipettes come in three or four standard sizes, designed to deliver less than 2 μL, 1–10 μL, 10–100 μL, or 100–1000 μL. The smallest size is usually not provided in teaching labs. A modern pipette works by air displacement: There is a piston that displaces a precise volume of air when the plunger is depressed, which displaces an equal volume of the liquid from the tip after the plunger is released. Only the disposable tip contacts the liquid. Most pipettes contain an automatic tip-ejector mechanism that is activated by pressing a button that activates a tip-ejector arm (Figure I.1). Tips should not be removed by twisting with the fingers because this can often take the piston assembly apart.

One set of micropipettes should be dispensed to each student team and the students instructed to label their set and

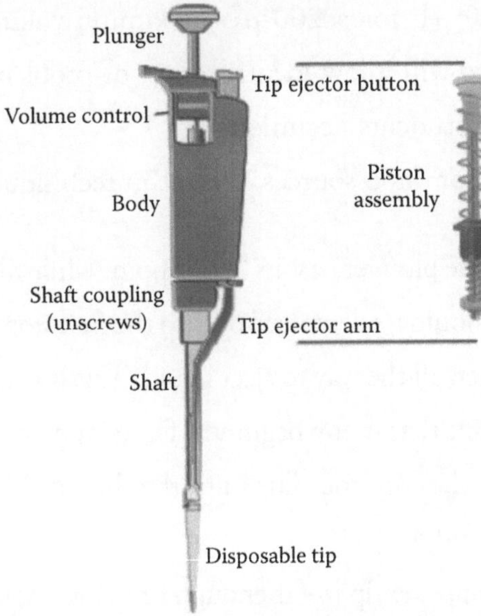

Figure I.1 Diagram of an air displacement micropipette. The volume control is adjusted within a set limit by a screw mounted on a micrometer. A disposable tip is placed firmly onto the end of the shaft. The plunger is depressed to its first stop, and the tip is placed in the desired liquid. The plunger is released, liquid is aspirated, and the liquid is ejected from the pipette by depressing the plunger all the way. The tip is then ejected into a waste container using the ejector button. The piston assembly mechanism is also shown.

inspect it for dirt or damage. Leakage can be assessed by filling a tip to its maximum volume and observing for several minutes to make sure the level does not drop. If it leaks, it must be repaired. The pipettes should then be calibrated by pipetting distilled water onto an analytical balance. A volume near the middle of the range of the tested pipette should be used (e.g., 100 μL for a 200-μL maximum-volume pipette). This procedure will allow identification of problems with the pipettes or the students' technique.

The most common sources of error in technique include:

- Pushing the plunger past its "stop" point while filling. The plunger should be depressed only to the first stop point for filling, then all the way to eject liquid. This is a somewhat-subtle point that many beginners fail to appreciate.
- Dipping the tip too far into the liquid (should be 2–3 mm only)
- Not dipping the tip in far enough or using a sample that is too shallow, thereby aspirating air
- Having a loose tip
- When pipetting viscous fluids such as glycerol, having a large amount of fluid stuck to the outside of the tip, which is delivered along with what is inside.

Anyone new to the use of micropipettes should practice until good technique is achieved before proceeding with any of the experiments in this book.

I.4 Suggested Schedules

Tables I.1 and I.2 provide suggested schedules for a semester-long and a summer intensive course, respectively.

Table I.1 Suggested Schedule for Semester-Long Course[a]

Week	Experiment
1	During lab period: Experiment 1A; next day: Experiment 1B
2	Experiment 2
3	Experiment 3 during lecture period; Experiment 4 during lab period
4	During lab period: Experiment 5A; next day: Experiment 5B
5	Experiment 6
6	During lecture hour: Experiment 7; during lab period: Experiment 8A
7	Experiment 8B (whenever results are ready)
8	Experiment 9 (Parts 1 and 2 by teaching assistant; Parts 3 and 4 done in class)
9	Experiment 10
10	Experiment 11 (examine crystals during lecture hour)
11	Experiments 12 and 13 (passaging done as demo)
12	Experiment 14 or Experiment 15
13	Experiment 16 or Experiment 17

[a] One 3-hour lab session per week with three course meetings allowing for brief experiments, such as colony picking, seeding, and the like.

Table I.2 Schedule for Intensive Course[a]

Week	Monday	Tuesday	Wednesday	Thursday	Friday	Weekend
1	Experiment 1, Parts 1–3 Also Experiment 9, Part 1 (save plates)	Lecture and introduction; Experiment 1, Part 4	Experiment 2, Parts 1–4	Experiment 3, Parts 1–3	Experiment 2, Part 5, and Experiment 3, Part 4 (one gel); Experiment 4, Parts 1 and 2	Ligation reactions run; instructor to transform them Sunday
2	Experiment 4, Part 3 Experiment 6: computer lab and Part 1	Experiment 5	Experiment 6, Parts 2–4 (may be moved if primers not available) Experiment 7, Part 1	Experiment 7, Part 2	Experiment 8	
3	Experiment 9 parts 3 and 4 (Parts 2 and 3 behind the scenes by teaching assistant)	Experiment 10	Experiment 11, Part 1	Experiment 11, Part 3	Experiment 12, Part 1 Experiment 8B	
4	Experiment 12, Part 2	Experiment 13	Experiment 14 as demo or Experiment 15	Experiment 16 or 17	Cleanup/catch up	

[a] Four 3-hour lab sessions per week.

Background Reading

Nadeau, J. *Introduction to Experimental Biophysics: Biological Methods for Physical Scientists*. Boca Raton, FL: Taylor & Francis; 2012, Chapter 1.

SECTION ONE

Molecular Cloning

1 Transformation of Competent Bacteria with a DNA Plasmid

1.1 Experiment 1A: Laboratory Setup and Transformation of Competent Bacteria with an Unknown Plasmid

1.1.1 Introduction

The bacterial strains you will use in this course have specific genetic properties that are important to the success of experiments. Ensuring that the bacterial strains have the proper phenotypic behavior will favor a successful outcome. Familiarize yourself with bacterial strains used for different purposes by consulting the references in the Background Reading section.

In the first laboratory session, teams are organized, and equipment used throughout the course is assigned to each team. Each team also selects a laboratory bench where experiments will be performed and a drawer for storage of materials and equipment. A team should use the same bench, drawer, and pipettes throughout the course. Materials for specific experiments will be made available on supply benches at the beginning of each session. Familiarize yourself with the location of refrigerators, freezers, incubators, and water baths: All will be needed in the first experiment.

The laboratory project begins with a simple experiment to transform a chemically competent bacterial strain with a "mystery plasmid." The resulting colonies will be used to generate DNA needed in subsequent experiments, in which you will figure out which plasmid you have and try to do something imaginative and interesting with it.

Your mystery plasmid may be resistant to either of two antibiotics: kanamycin (KanR) or ampicillin (ampR). You will prepare selective plates with both types of antibiotics and use this property as the first screening test in determining which of the mystery plasmids you have. See Figures 1.1 and 1.2 for the vector maps of the possible mystery plasmids.

1.1.2 Materials

1. Plasmid DNA at a specified concentration
2. Chemically competent cells
3. Electrically competent cells and electroporation cuvette (optional)
4. SOC medium (super optimal broth with catabolite repression)
5. Sterilized Bacto Agar and LB (Luria-Bertani) medium at 2X concentration
6. Petri dishes (pack of 24), aliquot of ampicillin, aliquot of kanamycin

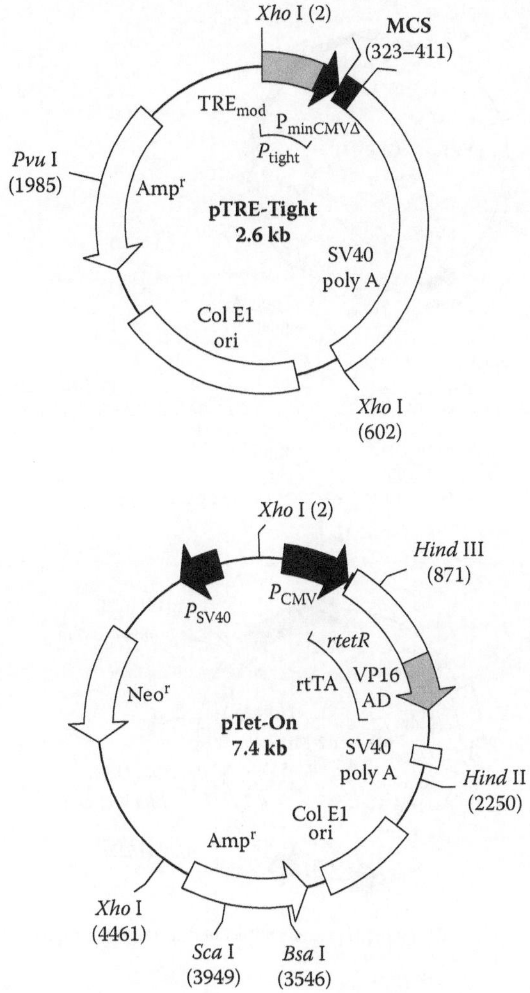

Figure 1.1 Possible identities of the mystery plasmids supplied in this exercise. These two plasmids are regulatory plasmids that confer tetracycline-inducible expression. Selected restriction sites are shown; ones in bold are unique (that is, occurring only once in the plasmid). (Images © Clontech Laboratories Inc., used with permission.)

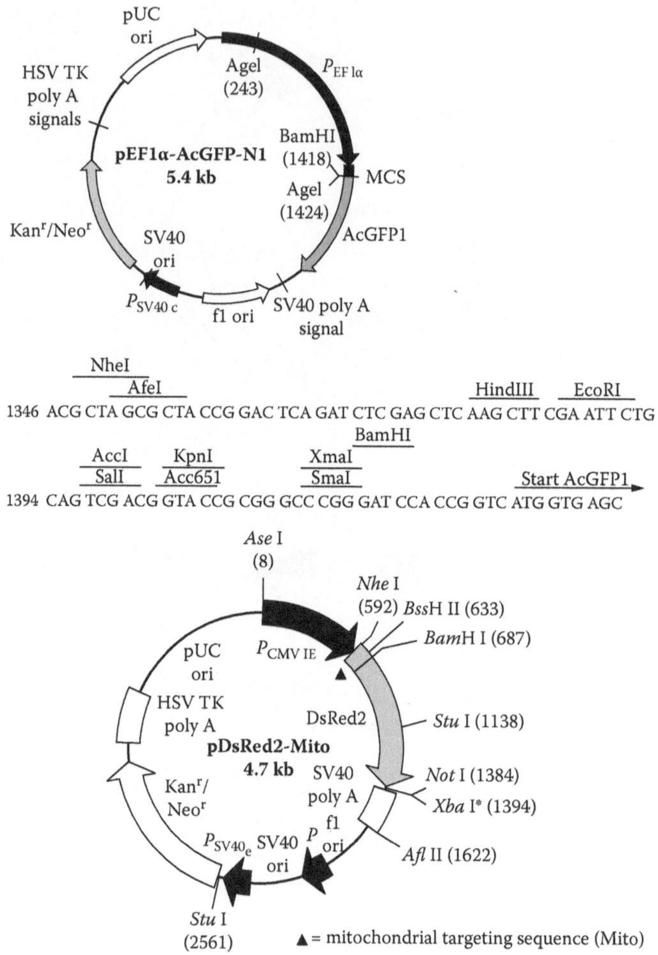

Figure 1.2 Other possible identities of the mystery plasmids. These plasmids serve as sources for fluorescent proteins, one green and one red. The red protein is targeted to cell mitochondria. A multiple cloning site (MCS) is shown as an example in one case; in both images, unique restriction sites are given in bold. (Images © Clontech Laboratories Inc., used with permission.)

7. Sharpies for labeling tubes
8. Ice bucket and ice
9. Sterile water

1.1.3 Experimental Procedures

1.1.3.1 Transformation of Chemically Competent Cells

1. Thaw on ice one vial of chemically competent cells.
2. Add 10 to 100 ng of DNA, in a volume of 1 to 5 μL, to the cells and mix by tapping gently. *Do not* mix cells by pipetting. They will die.
3. Incubate the vial(s) on ice for 30 minutes.
4. Incubate for exactly 30 seconds in the 42°C water bath. Do not mix or shake.
5. Remove vial(s) from the 42°C bath and quickly place on ice.
6. Add 250 μL of prewarmed SOC medium to the vial(s). **Note: SOC is a rich medium; practice good sterile technique to avoid contamination.**
7. Place the vial(s) in a microcentrifuge rack and secure with tape. Place the rack on its side in a shaking incubator and shake the vial(s) at 37°C for 1 hour at 225 rpm. While waiting, proceed to Section 1.1.3.3.
8. Plate 20 to 200 μL each of the transformation reaction onto four plates: two with ampicillin and two with kanamycin. Use different concentrations on each of the two identical plates to ensure well-spaced colonies.

9. Incubate the plates at 37°C overnight. **Make sure the plates are labeled with your name and the antibiotic. The teaching assistants will remove the plates from the incubator and store them in the 4°C refrigerator for you to use later.**

1.1.3.2 Transformation of Electrically Competent Cells (Optional)

If your instructor wishes you to use electrically competent cells, this part may be done with the same plasmids as provided in the previous part.

1. Remove an electroporation cuvette from the freezer and place on ice.
2. Thaw on ice one vial of electrically competent cells.
3. Add 5 to 10 ng of DNA, in a volume of 1 to 5 μL, to the cells and mix by tapping gently.
4. Transfer the DNA-cell mixture into the electroporation cuvette using a micropipette.
5. Follow the instructions on the electroporator.
6. After electroporation, immediately add 500 μL of pre-warmed SOC medium to the cuvette and pipette gently up and down several times.
7. Transfer the mixture to a 5-mL tube and place into the shaking incubator for 1 hour (while waiting, proceed to Section 1.1.3.3).

8. Plate 20 to 200 μL each of the transformation reaction onto two of the provided LB plates containing ampicillin and two plates containing kanamycin. Plate two different volumes to ensure well-spaced colonies on at least one plate.
9. Invert the plates and incubate at 37°C overnight.

1.1.3.3 Preparation of Antibiotic Plates

1. Place the 2X agar in the microwave until melted. **Caution: It may boil.**
2. Pour the 2X LB into the agar and swirl well to mix.
3. Place into the 60°C water bath until it has cooled (10–15 min).
4. Add the aliquot of antibiotic provided and swirl to mix.
5. Divide among your 24 plates, avoiding air bubbles.
6. Let air-dry on the bench and transfer into a plastic bag with your team names.

1.2 Experiment 1B: Selection of Colonies for Plasmid Screening

1.2.1 Introduction

Plated bacteria should be left in the 37°C incubator for 12–16 hours until well-defined, discrete colonies appear. Each of these contains a genetically identical plasmid "clone." If the plates are not allowed to grow long enough, the colonies will be too small to see.

Conversely, if they are left for too long, they will lose the plasmids you are trying to grow. *Escherichia coli* continuously exposed to ampicillin begins to excrete β-lactamase, which destroys the antibiotic and thus its ability to select for cells containing the plasmid of interest. On neglected ampicillin plates, you will see large colonies of bacteria surrounded by a halo of smaller colonies. These "satellite" colonies are growing in ampicillin-depleted areas and will almost certainly not contain a plasmid. A similar principle applies to kanamycin, although you are more likely to see a bunch of small colonies rather than satellites in this case.

1.2.2 Materials

1. Liquid LB medium
2. Ampicillin (100 mg/mL) or kanamycin (30 mg/mL)
3. Your plates from Part A

1.2.3 Experimental Procedures

1. Using sterile technique, pipette 5 mL of LB medium to each of five 15-mL tubes.
2. Check your plates to see which one had colonies. This is the one with the appropriate antibiotic resistance.
3. Add the appropriate antibiotic to four of the tubes to 100 μg/mL.
4. Take your plate that has the colonies and identify an area with single colonies. Choose one and "pick" it by touching it with a sterile micropipette tip, then

dropping the tip into the 15-mL tube. Repeat this with three more tubes. Leave one tube with LB medium only: This is a control for contamination of the medium, which should be performed routinely (meaning *always*).

5. Put the caps on loosely (to allow air circulation) and place the tubes in the shaker incubator.
6. The instructor will remove the tubes from the incubator after 8 hours and save them for the next class.
7. Wrap Parafilm® around the edges of all of your plates and store them at 4°C. You may need them later.

1.2.4 Expected Results and Pitfalls

The expected outcome of this exercise is the growth of 20–200 discrete, several-millimeter-size colonies on either the ampicillin or the kanamycin plates (but *not both*). This seems simple, but in practice, usually only about 50% of first attempts succeed. The hardest part of molecular biology is that it is hard to tell when something has failed right away. At least 12–16 hours need to pass with failure of cells to grow before it becomes obvious that an error has occurred. There are several ways in which bacterial transformation and plating can go wrong. In order of likelihood, these are:

- Killing the cells. Both chemically competent and electrically competent *E. coli* are very fragile. If the chemically competent cells were held in the water bath too long or they were shaken or the electrically competent cells were arced, then no colonies will grow on the plates. In the course, this is easily remedied by borrowing a colony or two from a neighboring team's plate, but you may wish to redo the transformation anyway, if possible, to improve your technique. Everyone kills competent cells sometimes—it takes practice to reduce this occurrence to a minimum.

- Forgetting to add antibiotic to the plates or adding it when the agar is too hot so that the antibiotic is destroyed. This is a surprisingly common error. In this case, the plates will be covered by bacterial growth, probably as a confluent "lawn" rather than discrete colonies. However, if poor technique (as previously discussed) is coupled with lack of antibiotic, the colonies can look great but contain no plasmid—you will not find this out for another few days. If both the ampicillin and kanamycin plates have colonies, this is a warning sign that something is wrong.
- Adding too much or too little DNA to the cells. Transformation efficiency will be greatly reduced in both of these cases. Watch your arithmetic and do not fall prey to a "more-is-better" attitude.
- Plating too much of the transformation solution. This can cause a lawn. Although bacteria in the lawn can grow, it is not good microbiological technique to pick

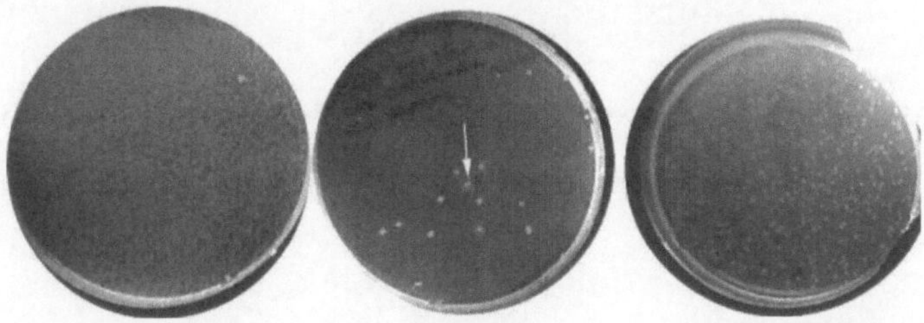

Figure 1.3 Appearance of colonies of *Escherichia coli* on an agar plate.

anything that is not a single colony because the cells will be genetically heterogeneous.

See Figure 1.3 for how plates should and should not appear.

Background Reading

Nadeau, J. *Introduction to Experimental Biophysics: Biological Methods for Physical Scientists.* Boca Raton, FL: Taylor & Francis; 2012, Sections 2.1–2.3.

Clontech: Vector guides and technical information for the Mystery Plasmids.

Froger, A. and J. E. Hall. Transformation of plasmid DNA into *E. coli* using the heat shock method. *J Vis Exp* (6):253 (2007).

Guides to *Escherichia coli* strains for different purposes:

New England BioLabs. Competent cells. http://www.neb.com/nebecomm/tech_reference/competent_cells/introductory.asp#.UMEJa4UTX6c

OpenWetWare. *E. coli* genotypes. http://openwetware.org/wiki/E._coli_genotypes

Sigma-Aldrich. Genotypes of selected *E. coli* strains. http://www.sigmaaldrich.com/life-science/molecular-biology/cloning-and-expression/learning-center/genotypes-of-e-coli.html

Questions

Please include the responses to the end-of-chapter questions in your laboratory notebook, with discussions or figures as necessary.

1. Did you obtain single colonies? Draw a diagram indicating what your plates looked like after overnight incubation at 37°C. (You should also provide a diagram of the plates in the Results section of your notebook.) If you failed to obtain single colonies, what would you do to improve your technique?
2. Can you estimate the transformation efficiency of the competent cells?
3. Was your plasmid ampicillin or kanamycin resistant?

2 Isolation of Plasmids by Miniprep and Restriction Enzyme Analysis

2.1 Experiment 2: Isolation of Plasmids by Miniprep and Characterization by Spectrophotometry

2.1.1 Introduction

The purpose of this lab is to extract the plasmid DNA from the transformed bacteria from Experiment 1 and purify it sufficiently to be used in molecular biological experiments. We will extract the DNA on a small scale, starting with about 5 mL of bacterial culture; this is called a *miniprep*. Larger volumes give rise to midipreps (25 mL of culture), maxipreps (200 mL), gigapreps, and so forth. To purify the plasmids, all other parts of the bacterial cell must be removed: cell walls and membranes, chromosomal DNA, RNA, and proteins. Each of the steps in this procedure is designed to successively remove these contaminants and yield pure plasmid DNA. The concentration and purity of the resulting DNA will then be estimated using absorbance spectroscopy in the ultraviolet (UV) range.

Although the spectra will not distinguish between plasmid DNA and contaminating nucleic acids (DNA or RNA), they will indicate the presence of excessive protein and provide a good indicator of the amount of DNA present.

2.1.2 Material

1. Cultures of transformed *Escherichia coli* prepared in Experiment 1
2. Microcentrifuge tubes
3. Tris-HCl
4. EDTA (ethylenediaminetetraacetic acid)
5. Glucose
6. Sodium dodecyl sulfate (SDS)
7. Sodium hydroxide concentrated solution
8. Potassium acetate (5 M, pH 4.8)
9. RNase A (in the freezer)
10. PCIA (phenol/chloroform/isoamyl alcohol) reagent
11. Isopropanol (kept cold)
12. Ammonium acetate (7.5 M)
13. DNA-grade absolute ethanol
14. Disposable spectrophotometry cuvettes

2.1.3 Experimental Procedures

2.1.3.1 Prepare Buffers

Using the reagents provided, prepare the following buffers using the lab pH meter or pH strips to adjust the pH:

1. GTE buffer: 50 mM glucose, 25 mM Tris-Cl, 10 mM EDTA, pH 8
2. TE buffer: 10 mM Tris-HCl, 1 mM EDTA, pH 7.5
3. Lysis buffer: 1% SDS in 0.2 M NaOH

2.1.3.2 Miniprep

1. For each miniprep, pipette 1.5 mL of the bacterial culture into a microcentrifuge tube. It is a good idea to do two or four minipreps in case something goes wrong (an even number helps balance the centrifuge). Spin the culture at top speed in a tabletop microcentrifuge for 60–90 seconds. Pipette off or dump out the supernatant; it is not a biohazard and may be poured down the drain. Make sure no excess medium remains.
2. If the bacterial pellet is very small (smaller than the end of the tube), repeat Step 1. This should not be necessary except if bacterial cultures were removed from the incubator too soon.
3. Resuspend the cell pellet in 100 μL of GTE buffer. Make sure the cells are well suspended. Do not worry about hurting them at this point.
4. Line up the miniprep tubes and add 200 μL of NaOH/SDS lysis solution. Take all the tubes in one hand and invert them half a dozen times. Make sure all tubes are handled at the same time, so you can proceed to Step 5.

5. **Immediately** add 150 μL of potassium acetate to neutralize the NaOH. The bacterial genomic DNA and SDS will form a thick, gooey precipitate.
6. Centrifuge at maximum speed for 60–90 seconds to settle the precipitate.
7. The plasmid DNA is still in the clear supernatant. Carefully transfer this into a fresh microcentrifuge tube, avoiding the white mass or any flecks.
8. Place the tubes on ice and add 0.5 mL of cold isopropanol. Incubate for 10 minutes.
9. Centrifuge at maximum speed for 60–90 seconds.
10. The plasmid DNA is now in the pellet. Pipette off the supernatant and discard it. Gently dissolve the pellet in 0.4 mL of TE buffer and add RNAse to 0.5 mg/mL.
11. Mix well and incubate at 37°C for 30 minutes to digest RNA.
12. Perform a PCIA extraction to remove proteins. Working in the fume hood, add about 300 μL of PCIA to each tube and mix using a vortex.
13. Centrifuge at maximum speed for 5 minutes.
14. The uppermost, aqueous phase contains the plasmid DNA. Transfer this layer to a fresh microcentrifuge tube.
15. Repeat the PCIA extraction (optional).
16. Place the tubes on ice. To precipitate the plasmid DNA, add 100 μL ammonium acetate and 1 mL ethanol and incubate for 10 minutes.

17. Centrifuge at maximum speed for 5 minutes.
18. The plasmid DNA pellet will be essentially impossible to see. Remove the ethanol carefully; once most of it is gone, you may invert the tube to allow the pellet to dry completely. Once all the ethanol is removed, dissolve the DNA in 50 μL TE buffer or ultrapure water.

2.1.3.3 Spectrophotometric Determination of Nucleic Acids

If all has gone well, the DNA in the minipreps is much too concentrated to read on the spectrophotometer. A small amount should be diluted 50–100 times before the measurement is taken.

1. Fill a spectrophotometry cuvette with water or TE buffer and run a blank.
2. Add the diluted DNA to the same cuvette. Run a measurement of absorbance at 260 and 280 nm or a full spectrum across the UV range (200–400 nm).
3. Calculate the concentration of DNA as

$$c = \frac{A_{260} - A_{blank}}{\varepsilon \ell}$$

   where ℓ is the path length of the cuvette, and ε is 6600 M^{-1} cm^{-1}.
4. Estimate the purity of the DNA by taking the ratio A_{260}/A_{280}.

2.1.4 Expected Results and Pitfalls

Minipreps are quite straightforward. An experienced molecular biology student might do hundreds of them in a day. It is a good idea to always do each one at least in duplicate, however, because accidental spills or bacteria lacking plasmids can always occur. Sometimes, a bacterial colony looks fine, gives a good pellet, but there is simply no plasmid. Most commonly, though, lack of plasmid DNA at the end of the miniprep is caused by handling errors. Allowing the lysis buffer to act for too long is probably the most common error—do not walk away and let it act before neutralizing it. Another common problem is dislodging the white precipitate after SDS precipitation and accidentally transferring a good deal of it into the "clean" tube. If this happens, spin and transfer again. It is also possible to accidentally take the wrong phase after PCIA extraction. Remember that the PCIA phase (**not** what you want) is on the bottom; the proteins are in a layer in the middle; and the aqueous phase (**is** what you want) is on the top (Figure 2.1a). Finally, it is important to remove all traces of ethanol in the last step or it will contaminate the DNA and may interfere with downstream applications. Often, a good deal is left behind by overcautious pipetting; in this case, it will take a long time to air-dry. Remove more with a low-volume micropipette, tilt the tube, or vacuum dry. A clean DNA prep will show a UV absorbance spectrum with a clear

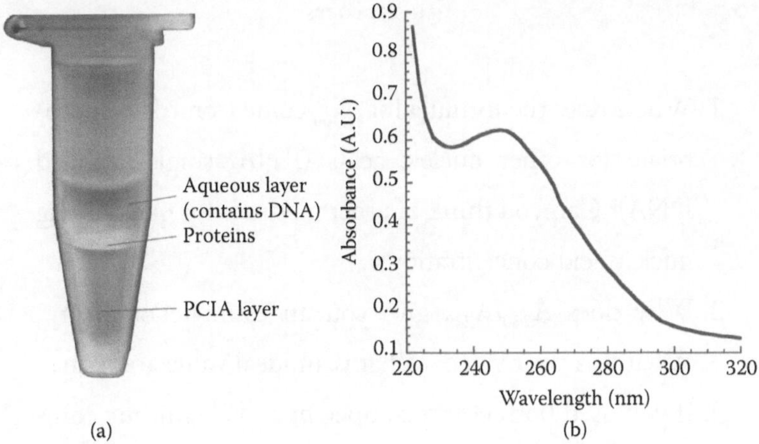

Figure 2.1 (a) Appearance of tube after PCIA addition and centrifugation. (b) Desirable absorbance spectrum for purified DNA.

peak at 260 nm (Figure 2.1b). If the location of the peak is off, check the spectrophotometer calibration and make sure to run a blank.

Background Reading

Nadeau, J. *Introduction to Experimental Biophysics: Biological Methods for Physical Scientists.* Boca Raton, FL: Taylor & Francis; 2012, Section 2.4.

Birnboim, H. C. and J. Doly. A rapid alkaline extraction procedure for screening recombinant plasmid DNA. *Nucleic Acids Res* 7, 1513–1523 (1979).

Questions

1. Where does the formula for A_{260} come from? Is it appropriate for other nucleic acids (RNA, single-stranded DNA)? Can you think of other methods for quantifying nucleic acid concentrations?
2. Why does A_{260}/A_{280} give you an estimate of purity? What was your value? What is an ideal value and why?
3. If you used both electrocompetent and chemically competent cells, which yielded more plasmid?

3 Designing a New Construct Using Restriction Enzymes (Computer Lab)

3.1 Experiment 3: Use of Software to Screen Plasmids and Develop Cloning Strategies (Computer Lab)

3.1.1 Introduction

This lab is a computer exercise designed to show how restriction enzymes can be used to cut plasmids at specific locations for screening or cloning purposes. It also will introduce you to how DNA sequences can be downloaded, entered, and mapped using different computer programs. The overall goals of this session are twofold: (a) to determine how the minipreps should be screened for identification of the mystery plasmid; and (b) to develop an idea for cut-and-paste cloning of the provided plasmids that can be implemented using restriction enzymes. Part b then forms the basis of the work in the next few exercises.

3.1.2 Materials

Computer with an Internet connection and cloning software (see Section 3.1.4 for more details) is needed.

3.1.3 Experimental Procedures

3.1.3.1 Obtain Sequences and Import into Software

1. From the manufacturer's website, obtain the sequence of each of the mystery plasmids in text format.
2. Input the sequence into the plasmid-mapping software installed on the lab computer.
3. Practice different styles of plasmid mapping. Draw the plasmid as a circle, indicating key regions (gene of interest, promoters, multiple cloning sites).
4. Indicate restriction enzyme sites on the plasmid. Experiment with the most useful labeling: six hitters only, unique sites, and so on.
5. Print your final maps and paste them into your lab notebook.

3.1.3.2 Develop a Screening Plan

1. Using the provided software, figure out how to simulate gel electrophoresis of plasmids cut with selected restriction enzymes.
2. Choose enzymes that you predict will allow distinction of the mystery plasmids. Try not to cut any plasmid with

more than two enzymes at once; however, more than one digest per plasmid may be desirable.

3. Simulate a gel that will distinguish all four plasmids. Print it and save it.

3.1.3.3 Develop a Cloning Plan

1. Think of ways in which the mystery plasmids may be cut and pasted to give something new and interesting.
2. Come up with a "cloning plan" for how this can be accomplished using restriction enzymes.
3. Simulate gels for each of the steps in the cloning process. Indicate which fragments will need to be gel purified. Print the steps and save them; they will come in handy for the next session.

3.1.4 Software Suggestions

3.1.4.1 Commercial Software

Commercial software are often costly programs that the instructor's laboratory may or may not be able to provide. They are full-service packages that do everything: uploading sequences, mapping, optimizing codons, analyzing open reading frames (ORFs), and more.

1. DNASTAR®'s SeqBuilder (http://www.dnastar.com/t-sub-products-lasergene-seqbuilder.aspx). Part of the DNASTAR family of molecular biology tools, SeqBuilder

does plasmid mapping, annotation, and virtual cloning with automatic cloning planning. Personal computer (PC) and Mac versions are available.

2. Gene Construction Kit® (http://www.textco.com/): full-featured mapping program that allows tracking of cloning project history. Mac and PC forms are available with easy cross-platform compatibility.
3. Vector NTI® (http://www.lifetechnologies.com/us/en/home/life-science/cloning/vector-nti-software.html): full-featured package from Invitrogen; available for both PC and Mac. Both Advance and Express packages are available at different prices. Molecular biologists frequently exchange files saved in Vector NTI format, and companies selling plasmids often give their sequences in this format.
4. SimVector (http://www.premierbiosoft.com/plasmid_maps/index.html). Mapping, virtual cloning, and more. Both PC and Mac versions are available; it can open Vector NTI files.
5. CloneMap (PC) and MacPlasmap (Mac) (http://www.macgui.com/downloads/?file_id=15727). Relatively low-cost, multifunctional virtual cloning software.
6. BVTech Plasmid (http://www.biovisualtech.com/bvplasmid/downloadpl.htm). Low-cost package with a student discount (for PC only). It contains many nice

features, such as custom mapping, import and export of different file types, multilingual support, and more.

7. NoeClone (http://www.noegen.com/en/product_clone.php). Three levels of molecular biology software are available: Basic (mapping only), Advanced (virtual gels and cloning plans), and Enterprise (high-throughput cloning and database management) (PC and Mac).

3.1.4.2 Freeware

1. XPlasMap (http://www.iayork.com/XPlasMap). For the Mac. Draws and annotates plasmid maps and permits cut-and-paste "virtual cloning."
2. pDRAW32 (http://www.acaclone.com/). For Windows (or Mac running Softwindows or other emulator). Currently this is freeware, but licenses are available that may provide benefits. It provides mapping, annotation, virtual gels, and other features.
3. ApE (A plasmid Editor) (http://biologylabs.utah.edu/jorgensen/wayned/ape/). Mapping and sequence analysis tools with many useful features (Mac and PC). This is freeware with voluntary donation suggested.
4. PlasmaDNA (PLASmid MAnipulator) (http://research.med.helsinki.fi/plasmadna/). Very nice multifeature program that allows for cloning plan development and virtual gels (Mac and PC).

3.1.4.3 Online Tools

The programs mentioned in this section are platform independent and can be used online. All of them are highly recommended and used in my laboratory every day.

1. NEBcutter V2.09 (http://tools.neb.com/NEBcutter2). Plasmid mapping (from New England BioLabs).
2. PlasMapper (http://wishart.biology.ualberta.ca/PlasMapper). An online tool from the University of Alberta that generates high-quality circular maps and text maps from sequences pasted in or uploaded from vendor sites.
3. EZ-Plasmid (http://www.infosake.com/plasmid/index.html). Allows cut-and-paste import of sequences; will find genes, tags, restriction sites, and other features.
4. Analyze Sequence (http://www.addgene.org/analyze-sequence/). Tool from addgene. Paste in a sequence and get a map, restriction digest, ORF analysis, translation, and more.

3.1.5 Expected Results and Pitfalls

It will take some time and practice to become used to plasmid-mapping software, and preferences vary widely for mapping tools. Some of the links in this chapter may be outdated, and new tools are appearing all the time. Take your time to familiarize yourself with the concepts and operations and with finding the key parts of a plasmid in a map: the ORFs, promoters, start and stop codons, and so on. Try not to make a cloning scheme that is too elaborate or that makes use of more than one blunt-ended cutter; recall that blunt-ended ligations are much less efficient than sticky-ended ones.

Figure 3.1 shows an example of a plasmid map and virtual gel made with the freeware PlasmaDNA for Mac.

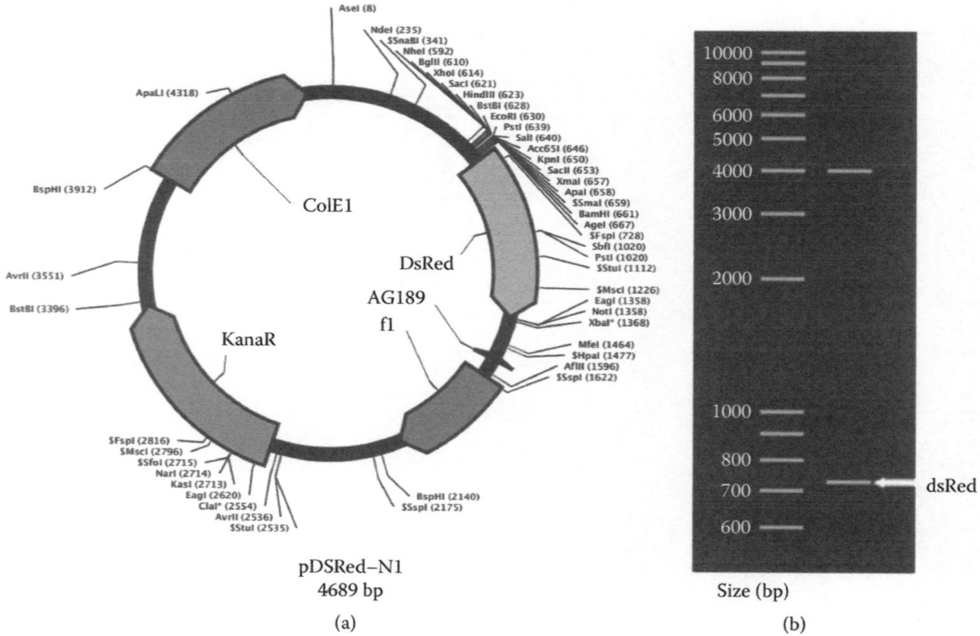

Figure 3.1 Plasmid mapping of Clontech plasmid pDsRed2-N1 using PlasmaDNA: (a) restriction map; (b) virtual gel after cutting with selected enzymes as indicated.

Background Reading

Nadeau, J. *Introduction to Experimental Biophysics: Biological Methods for Physical Scientists*. Boca Raton, FL: Taylor & Francis; 2012, Sections 2.5, 2.6.

Boyer, H. W. DNA restriction and modification mechanisms in bacteria. *Annu Rev Microbiol* 25:153–176 (1971).

Questions

1. Most restriction enzymes are "6-hitters"; that is, they recognize a specific sequence of six base pairs. However, 4-hitters are also common, and 8- and 10-hitters also exist. When would 8- or 10-hitters be needed? Why are 4-hitters not commonly used for cloning?
2. Explain the difference between a 3′ overhang, a 5′ overhang, and a blunt end.
3. Describe the types of degeneracy that can occur with restriction enzyme sequences.
4. Sometimes, the sticky end of one enzyme will religate with the sticky end of another, even though they do not recognize the same sequence. Give an example of this and explain how it can work. Which enzyme, if any, will recut the ligated piece?
5. When using blunt ends or a single restriction site, how is directionality established?

4 Restriction Enzymes

4.1 Experiment 4: Restriction Enzymes for Screening and Cloning

4.1.1 Introduction

This laboratory begins with restriction enzyme digestion and agarose gel analysis of the plasmids from Experiment 2. This analysis is performed to identify which of the mystery plasmids each team has. The teams will then share their plasmids with others to permit implementation of the various cloning strategies designed in Experiment 3.

4.1.2 Materials

1. Minipreps from Experiment 2
2. Restriction enzymes and buffers
3. Electrophoresis buffer (tris-acetate-EDTA [TAE]), 10X concentration
4. Agarose
5. Ethidium bromide

6. Ice buckets or "cold cups" for holding restriction enzymes
7. Gel tray and comb; lab tape
8. Gel loading buffer "6X" (50% v/v glycerol and 0.5% w/v bromophenol blue)
9. DNA ladder(s)

4.1.3 Experimental Procedures

4.1.3.1 Perform Small-Scale Restriction Digest to Identify Mystery Plasmid

1. Identify the enzymes to be used for screening from Experiment 3.
2. Find the enzymes in the –20°C freezer. Hold them in a cold cup or on ice in an ice bucket at all times. *Do not* allow the restriction enzymes to come to room temperature.
3. Locate the appropriate buffer for your enzyme(s). If you are using more than one restriction enzyme at a time, determine which buffer is optimal for using both online tools and the manufacturer's catalog. Thaw the buffer thoroughly; make sure any precipitate is redissolved before using.
4. Determine what volume of your miniprep DNA contains about 500 ng of nucleic acid (call it x).
5. Add 15 – 1.5 – 1 – x μL of distilled water (dH_2O) to a microcentrifuge tube.
6. Add 1.5 μL of restriction enzyme buffer to the tube.

7. Add your volume of DNA.
8. Add 1 μL of enzyme (if only one) or 0.5 μL each if there are two. Note that the enzymes contain glycerol, so the solution is viscous; be careful not to pipette too much.
9. Always follow that order: water, buffer, DNA, enzyme. This ensures adequate volumes.
10. Place into the appropriate water bath for 1 hour.

4.1.3.2 Gel Electrophoresis

1. Measure 50 mL of TAE buffer into a 200-mL or larger Erlenmeyer flask.
2. Add 0.8% agarose by weight.
3. Melt the agarose in the microwave oven. Be careful during this step or the agarose will boil over. Microwave for 1 minute, then hold the flask (using heat-protective gloves) and swirl it gently. Then, microwave for another 30 seconds.
4. When the agarose is fully melted, place the flask into a 60°C water bath to cool. This will prevent the agarose from solidifying.
5. Prepare a gel tray for casting. Several teams may share a single gel. Consult the instructor about casting the gel and prepare the tray.

6. Take your cooled agarose from the water bath, add 1.5 μL ethidium bromide (*danger*: carcinogen) to the agarose, swirl, and pour it into the tray.
7. When the gel has solidified, place it into the electrophoresis tank and fill the tank with TAE until the gel is just submerged.
8. To increase specific gravity and visibility of the samples, add 6X dye to all of your samples and to an appropriate amount of DNA ladder.
9. Apply the samples carefully to the agarose gel wells. Record the order of the samples and their respective wells.
10. Add 15 μL of ethidium bromide to the positive pole (bottom of tank).
11. Electrophorese the samples at 100 V (constant voltage) for 45–60 minutes. When done, let the instructor know.
12. Remove the gel carefully from the tank and take it to the gel imaging area.
13. Take a photo of your gel and print a copy for each member of your team.

4.1.3.3 Identify and Share Plasmids

1. Which mystery plasmid did you have?
2. Identify which teams had which plasmid.
3. Share all four mystery plasmids so that each team has some of each.

4.1.3.4 Large-Scale Digest to Obtain Fragments for Cloning Plan

1. Select the plasmids needed for the cloning plan from Experiment 3 and the enzymes needed to digest them.
2. Determine what volume of the plasmids contains about 3 μg of nucleic acid (*x*).
3. Add $50 - 5 - 3 - x$ μL of dH_2O to each of two microcentrifuge tubes, one for the vector and one for the insert.
4. Add 5 μL of the appropriate buffer to the tube.
5. Add your volume of DNA.
6. Add 1.5 μL of each enzyme.
7. Place into the appropriate water bath.

The instructor will remove the samples from the water baths after 2–4 hours and save them at -20°C for the next session.

4.1.4 Expected Results and Pitfalls

It is probably safe to say that nearly everyone makes a mess of his or her first agarose gel. This often literally means a mess: If the gel tray is not prepared properly or the agarose is too hot, the molten agarose will flow out of the gel tray and all over the benchtop. If this happens, do not touch it. Let it cool, then scoop it up and dispose of it with the ethidium bromide solid waste because it contains the dye. **Never** put agarose down a sink—it will clog the plumbing.

A few things can go wrong once the gel is prepared. Getting the samples into the wells is challenging at first because the gel is nearly transparent in the TAE buffer. Take your time and make sure to expel the sample near the bottom of the well. If you are having difficulties and have empty wells, practice in a few wells with the DNA ladder until your technique improves.

Another common error is to reverse the polarity of the gel box. If the DNA starts to migrate the wrong way, simply shut the box off, flip the cables, and start again. It is not ruined unless the DNA has gone so far that it runs off the top of the gel. Similarly, do not let the gel run too long, or the smaller fragments will run off the bottom and become invisible. Note the position of the loading dye with respect to the bottom of the DNA ladder; it should be about two-thirds of the way down when the gel is shut off.

Once the gel is run, it will need to be interpreted. If you have chosen your enzymes judiciously, there should be no problem telling the different mystery plasmids apart. However, bear in mind two things: (a) Agarose gels distinguish DNA fragments by size. Two fragments of the same size that are otherwise completely nonidentical will look alike. (b) We are using a 0.8% agarose gel, which can separate fragments from a couple hundred base pairs to a few kilobases in length. Any fragments smaller than that will run off the gel. Larger fragments will hardly migrate and so be essentially indistinguishable.

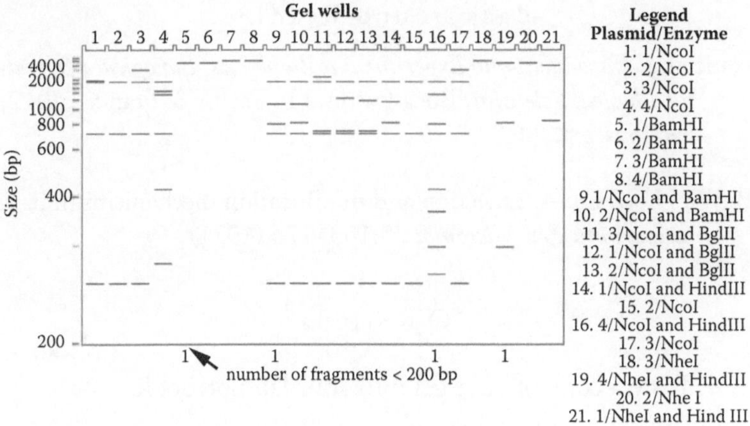

Figure 4.1 Simulated gel for distinguishing four distinct plasmids. Real gels will probably not have this many wells, so multiple gels would need to be run to obtain all of these results. This is OK if there are multiple teams using different enzymes. Things to note: (1) NcoI will not distinguish Plasmids 1 and 2 (the darker band in Plasmid 1 means there are two fragments of nearly equal length at about 2000 base pairs); (2) NcoI will distinguish Plasmids 3 and 4, so anyone with either of these would know it right away using this enzyme alone; (3) BamHI alone and BamHI with NcoI both also do a poor job of distinguishing Plasmids 1 and 2; (4) NcoI and BglII might do the job, but it is iffy; (5) NheI and HindIII can distinguish Plasmids 1 and 2; (6) sometimes, there are small fragments that run off the bottom of the gel, which are indicated by cloning software but which you cannot see in a real gel.

Figure 4.1 shows an example of a virtual gel designed to distinguish four mystery plasmids, here called Plasmids 1, 2, 3, and 4.

Background Reading

Nadeau, J. *Introduction to Experimental Biophysics: Biological Methods for Physical Scientists.* Boca Raton, FL: Taylor & Francis; 2012, Sections 2.5, 2.6

Boyer, H. W. DNA restriction and modification mechanisms in bacteria. *Annu Rev Microbiol* 25:153–176 (1971).

Questions

1. Paste a copy of your gel into your lab notebook.
2. Were you able to identify your plasmid unambiguously?
3. Did any ambiguity occur in the class? Discuss.

5 Ligation

5.1 Experiment 5A: Gel Purification and Ligation

5.1.1 Introduction

In this session, we will run the large-scale digests from Experiment 4 on an agarose gel. Rather than just take a picture, however, this time we will physically remove the pieces and purify them from the agarose. We will then use these purified pieces in a ligation reaction to attempt to stick them together. The efficiency of DNA ligase is extremely small, so the resulting ligation product is not useful as is. Instead, we will transform the ligation reaction into highly competent *Escherichia coli*. If things go right, the bacteria will amplify only the desired ligation product and not any undesirable pieces, such as unrestricted vector, recircularized vector, and so on. The colonies will grow overnight and will be picked by students or instructors the following day.

5.1.2 Materials

1. Large-scale digests from Experiment 4
2. Electrophoresis buffer (TAE), 10X concentration
3. Agarose
4. Ethidium bromide
5. Gel tray and comb; lab tape
6. Gel loading buffer 6X (50% v/v glycerol and 0.5% w/v bromophenol blue)
7. DNA ladder(s)
8. Face shield
9. Razor blades or plastic knives
10. Gel extraction kit
11. Ten-minute DNA ligase and ligase buffer
12. Cold cups or ice buckets
13. Pellet Paint®, sodium acetate, absolute ethanol (optional)
14. Ultracompetent *E. coli*
15. Ampicillin and kanamycin plates from Experiment 1

5.1.3 Experimental Procedures

5.1.3.1 Gel Electrophoresis with Fragment Extraction

1. Measure 50 mL of TAE buffer into a 200-mL or larger Erlenmeyer flask.
2. Add 0.8% agarose by weight.
3. Melt the agarose in the microwave oven. Be careful during this step or the agarose will boil over. Microwave

for 1 minute, then hold the flask (using heat-protective gloves) and swirl it gently. Then, microwave for another 30 seconds.

4. When the agarose is fully melted, place the flask into a 60°C water bath to cool. This will prevent the agarose from solidifying.
5. Prepare a gel tray for casting. Unlike in Experiment 4, this time you will need large wells for loading the whole digest of 3 μg of DNA. Use a comb with large teeth or "create" large teeth by placing tape across two or three teeth on a normal comb (Figure 5.1).
6. Take your cooled agarose from the water bath, add 1.5 μL ethidium bromide (danger: carcinogen) to the agarose, swirl, and pour it into the tray.

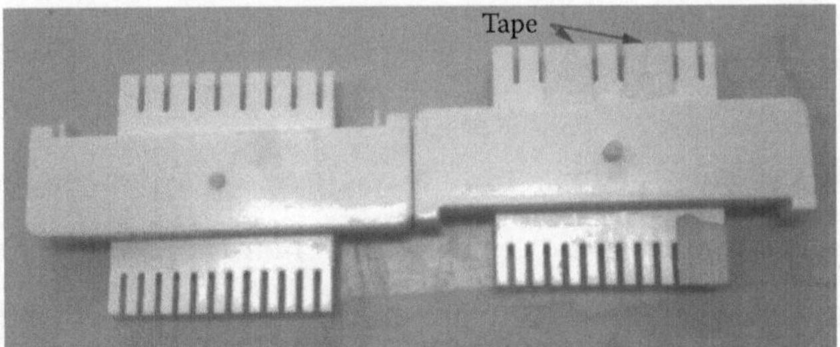

Figure 5.1 A gel comb and how to make the teeth wider with tape. This will create a single deep well for loading of a large sample. This is easier to load, cut out, and purify than several small wells.

7. When the gel has solidified, place it into the electrophoresis tank and fill the tank with TAE until the gel is just submerged.
8. To increase specific gravity and visibility of the samples, add 6X dye to all of your samples and to an appropriate amount of DNA ladder.
9. Apply the samples carefully to the agarose gel wells. Record the order of the samples and their respective wells.
10. Add 15 μL of ethidium bromide to the positive pole (bottom of tank).
11. Electrophorese the samples at 100 V (constant voltage) for 30 minutes or so, just enough to separate the fragments.
12. Remove the gel carefully from the tank and take it to the gel imaging area along with two razor blades or plastic knives and an ultraviolet (UV)-protective face mask (also wear goggles).
13. Take a photo of your gel and print a copy for each member of your team. Make sure you see the expected fragments.
14. If you do see the expected fragments, put on the face shield and look directly at your gel on the UV imaging tray. Using one razor blade or plastic knife, cut out the bright area corresponding to the "vector" in your cloning plan. If you see unexpected bands, consult the instructor. Do not remove too much agarose or you will have difficulty purifying the DNA.

15. Using the second razor blade, cut out the band corresponding to your "insert." Be careful not to forget which razor blade is which.
16. Take the photos and the extracted samples back into the lab when you are finished.

5.1.3.2 Gel Extraction

1. Tare an Eppendorf tube.
2. Transfer the extracted vector fragment to the tube and weigh it. Note its weight in your notebook.
3. Do the same with the insert fragment.
4. Follow the instructions in the provided gel extraction kit to purify the DNA from each fragment.
5. Estimate the concentration of the purified DNA based on the initial amount run, the relative molecular weight of the extracted piece, and the estimated amount lost in purification. The gel purification kit can be expected to be about 80% efficient barring any accidents (piece too big to purify, some dropped on the floor, etc).

5.1.3.3 Ligation

1. Once you have concentrations of vector and insert, you are ready to ligate.
2. Determine the volumes needed for a total of 100–200 ng of DNA at a 3:1 insert/vector *molar* ratio (total volume of x).

3. Obtain the ligase and ligation buffer. Keep the ligase on ice or in a cold cup. Thaw the buffer *completely* and mix it well.
4. Add 20 – x – 4 μL of H_2O to a microcentrifuge tube.
5. Add 2 μL of ligation buffer.
6. Add the DNA.
7. Add 2 μL T4 DNA ligase.
8. Prepare **2 controls**: one without insert and one without insert or ligase.
9. These values are based on 20 μL total volume for ligation. This is a good volume because the entire reaction can be put into a vial of *E. coli*. However, if your DNA is too dilute to obtain 100–200 ng in this volume, use a 50-μL reaction instead, with buffer scaled accordingly.
10. Let the reaction proceed for the recommended amount of time. For 10-minute ligase, this is just 10 minutes.

5.1.3.4 Cleaning the Ligation with Pellet Paint (Optional)

This optional step gets rid of enzymes, buffer ingredients, and extra water that may reduce the transformation efficiency in the next step. It is highly recommended—for some types of ligase, it is necessary. Read the package insert to be sure. The ethanol precipitates the DNA; the "paint" only serves to make it visible. Because of the tiny volumes of DNA used in a ligation reaction, it really helps to have this visual clue to the location of the DNA pellet.

1. To each tube (ligation and controls), add 2 μL of Pellet Paint, mix briefly, then add 2 μL of sodium acetate and mix again.
2. Add 50 μL of ethanol to each tube, vortex, and let incubate at room temperature for 5 minutes.
3. Spin at maximum speed in a microcentrifuge for 5 minutes. You should see a tiny spot the color of the Pellet Paint (blue or pink). This is your DNA. Remove the supernatant with a micropipettor.
4. Let the pellet dry briefly and resuspend in about 10 μL of H_2O. The DNA is resuspended when the paint goes into solution.

5.1.3.5 Transformation into Ultracompetent E. coli

1. Transform electrically or chemically competent *E. coli* as in Experiment 1.
2. Use the entire volume of the ligation if purified using Pellet Paint. Otherwise use half of it (10 μL).
3. After incubating the cells for 1 hour in SOC, plate them on ampicillin or kanamycin plates. While waiting, proceed to Step 5.1.3.6.
4. You will have three plates: a ligation reaction; a control without insert; and a control without insert or ligase.

5.1.3.6 Planning How to Screen the Transformants

While the bacteria are recovering from the shock of transformation, give some thought to how you will tell if a ligation worked. Ligation products will be miniprepped as in Experiment 2 and then screened by restriction enzyme digestion. The most obvious way to screen to see if the ligation worked is to cut the resulting plasmid with the same enzymes used for cloning in the fragment. However, there are caveats. If one of the enzymes used was a blunt cutter, it may not religate. Also, if the piece removed from the original vector was the same size as the piece inserted in the ligation, how would you tell uncut original vector from the correct ligated product? If a computer is available, run some virtual gels to determine a screening strategy.

5.2 Experiment 5B: Selection of Colonies for Ligation Screening

5.2.1 Introduction

Unlike in Experiment 1, the colonies that grow on ligation plates may not all be desirable. Some contaminants representing undigested vector or just plain "bad" ligations are likely to be present. There are two ways in which the ligation could have gone wrong: There could be no colonies on the experimental plate; or there could be colonies on one or both of the control plates. Because of the huge amount of DNA that was transformed, if there was any supercoiled DNA present, the control plates will likely show a confluent bacterial "lawn" rather than individual colonies. If you see this, do not bother to take any bacteria from the ligation plate; it also will be swamped with negative colonies.

Ideally, the experimental plate will show 10–100 colonies, and the control plates very few (<5) or none at all. If this is the case, go ahead with the experiment. If it is not, ask the teaching assistants to provide you with a plate or borrow one from another team.

Even when all seems well, most molecular biologists pick 10–12 colonies to screen from a ligation reaction in case a few of them are bad.

5.2.2 Materials

1. Liquid LB (Luria-Bertani) medium
2. Ampicillin (100 mg/mL) or kanamycin (30 mg/mL)
3. Plates from Experiment 5A

5.2.3 Experimental Procedures

1. Using sterile technique, pipette 5 mL of LB medium to each of ten to twelve 15-mL tubes.
2. Add the appropriate antibiotic to each of the tubes to 100 μg/mL.
3. Pick colonies from the ligation plates.
4. Put the caps on loosely (to allow air circulation) and place the tubes in the shaker incubator.
5. The instructor will remove the tubes from the incubator after 8–16 hours and save them for the next class.

5.2.4 Expected Results and Pitfalls

Ligations are tricky, and they usually just do not work the first time for any number of reasons: not enough vector or insert or total DNA in the reaction; loss of reaction while purifying; dead ligase; incomplete digestion of the vector (hard to tell when two enzymes cut close together); or simply *E. coli* that are not quite competent enough. Sometimes, the control plates will help pinpoint what went wrong. Other times, there are simply no colonies on the ligation or the controls (this is what happens with dead ligase or lost DNA), or the plates give garbage results despite looking great. Do not despair. In the real world of molecular biology, you would simply repeat the ligation until it works, changing the amount of DNA or other variables each time. This is the type of troubleshooting that requires deep thought and experience; hopefully this class will give you some idea of how nontrivial molecular biological manipulations can be.

There are ways to troubleshoot all of these problems, but for purposes of this class, if your ligation did not work, borrow some colonies from someone else's plate to have DNA for Experiment 6.

Background Reading

Nadeau, J. *Introduction to Experimental Biophysics: Biological Methods for Physical Scientists*. Boca Raton, FL: Taylor & Francis; 2012, Sections 2.5, 2.6.

Lehman, I. R. DNA ligase: structure, mechanism, and function. *Science* 186:790–797 (1974).

Questions

1. Can you estimate the minimum amount of DNA that can be visualized on an agarose gel?
2. Paste in a copy of your gel and estimate the fluorescence intensity of the different bands relative to the DNA ladder.
3. Did you see any bands other than those expected? If so, did they spoil your experiment?
4. What can go wrong in a ligation? What is the significance of the various controls?
5. Why do we use a 3:1 insert/vector ratio in a ligation reaction?
6. What possible products could we obtain other than the desired one? Can you estimate their probability?
7. How would you screen for dead ligase? How about for lost DNA?
8. For Experiment 5B, draw your three plates in your lab notebook. Do you think your ligation worked?

6 Screening a Ligation and Choosing Recombinants

6.1 Experiment 6: Screening Recombinants by Miniprep and Restriction Analysis

6.1.1 Introduction to Experiment 6

This exercise makes use of all you have learned in the previous sessions. Here, you will miniprep all of the colonies picked from your ligation plate and screen the resulting DNA using restriction enzyme digestion and agarose gel analysis. By now, these procedures will begin to feel routine; what is different here is that we are not sure of the end product; remember, many things can go wrong with ligations. If restriction analysis with one enzyme gives ambiguous results, you may need to use another—this can be slow. Judicious choice of enzymes and cleverness in interpreting the results are necessary when doing ligations.

6.1.2 Materials Provided for Experiment 6

1. Seeded colonies from Experiment 5
2. Microcentrifuge tubes
3. GTE buffer: 50 mM glucose, 25 mM Tris, 10 mM EDTA (ethylenediaminetetraacetic acid), pH 8
4. TE buffer: 10 mM Tris, 1 mM EDTA, pH 7.5
5. Lysis buffer: 1% sodium dodecyl sulfate (SDS) in 0.2 M NaOH
6. Potassium acetate (5 M, pH 4.8)
7. RNase A (in the freezer)
8. Isopropanol (kept cold)
9. Ammonium acetate (7.5 M)
10. PCIA (phenol/chloroform/isoamyl alcohol) reagent
11. Isopropanol (kept cold)
12. Ammonium acetate (7.5 M)
13. DNA-grade absolute ethanol
14. Restriction enzymes and buffers
15. Electrophoresis buffer (tris-acetate-EDTA [TAE]), 10X concentration
16. Agarose
17. Ethidium bromide
18. Gel tray and comb; lab tape
19. Gel loading buffer "6X" (50% v/v glycerol and 0.5% w/v bromophenol blue)
20. DNA ladder(s)

6.1.3 Experimental Procedures

6.1.3.1 Many, Many Minipreps

1. For each miniprep, pipette 1.5 mL of the bacterial culture into a microcentrifuge tube. Spin the culture at top speed in a tabletop microcentrifuge for 60–90 seconds. Pipette off or dump out the supernatant, making sure no excess medium remains.
2. If the bacterial pellet is very small (smaller than the end of the tube), repeat Step 1. This is sometimes necessary with ligations, although it may indicate an undesirable product.
3. Resuspend the cell pellet in 100 μL of GTE buffer.
4. Line up the miniprep tubes and add 200 μL of NaOH/SDS lysis solution. Take all the tubes in one hand and invert them half a dozen times. Make sure all tubes are handled at the same time, so you can do the next step immediately.
5. **Immediately** add 150 μL of potassium acetate to neutralize the NaOH. The bacterial genomic DNA and SDS will form a thick, gooey precipitate.
6. Centrifuge at maximum speed for 60–90 seconds to settle the precipitate.
7. The plasmid DNA is still in the clear supernatant. Carefully transfer this into a fresh microcentrifuge tube, avoiding the white mass or any flecks.

8. Place the tubes on ice and add 0.5 mL of cold isopropanol. Incubate for 10 minutes.
9. Centrifuge at maximum speed for 60–90 seconds.
10. The plasmid DNA is now in the pellet. Pipette off the supernatant and discard it. Gently dissolve the pellet in 0.4 mL of TE buffer and add RNase to 0.5 mg/mL.
11. Mix well and incubate at 37°C for 30 minutes to digest RNA.
12. Perform a PCIA extraction to remove proteins. Working in the fume hood, add about 300 μL of PCIA to each tube and mix using a vortex.
13. Centrifuge at maximum speed for 5 minutes.
14. The uppermost, aqueous phase contains the plasmid DNA. Transfer this layer to a fresh microcentrifuge tube.
15. Place the tubes on ice. To precipitate the plasmid DNA, add 100 μL ammonium acetate and 1 mL ethanol and incubate for 10 minutes.
16. Centrifuge at maximum speed for 5 minutes.
17. Remove the ethanol carefully and allow the pellet to dry completely. Once all the ethanol is removed, dissolve the DNA in 50 μL ultrapure water. Do not use any other buffers because restriction buffers will be added in further steps.
18. Label the miniprep tubes carefully. It is easy to mix up this many samples. Give them a simple name or number and be careful not to mix them up.

6.1.3.2 Restriction Digestion of Minipreps

1. Locate the restriction enzyme(s) you have identified as best for screening in Experiment 5.
2. Prepare a "master mix" of water, buffer, and enzyme, enough for 11 reactions if you have done 10 minipreps. This is 126.5 μL water; 16.5 μL buffer; μL total enzyme (5.5 μL of each if using two enzymes).
3. Pipette 13 μL of this reaction into each of 10 tubes.
4. Add 2 μL of miniprep to each tube—make sure you label carefully.
5. Place the tubes in the water bath for 1 hour. Immediately proceed to the next part.

6.1.3.3 Agarose Gel Electrophoresis of Digested Products

1. Measure 50 mL of TAE buffer into a 200-mL or larger Erlenmeyer flask.
2. Add 0.8% agarose by weight.
3. Melt the agarose in the microwave oven. Be careful during this step or the agarose will boil over. Microwave for 1 minute, then hold the flask (using heat-protective gloves) and swirl it gently. Then, microwave for another 30 seconds.
4. When the agarose is fully melted, place the flask into a 60°C water bath to cool. This will prevent the agarose from solidifying.

5. Prepare a gel tray for casting. Use a comb with small teeth and at least 11 wells.
6. Take your cooled agarose from the water bath, add 1.5 μL ethidium bromide (danger: carcinogen) to the agarose, swirl, and pour it into the tray.
7. When the gel has solidified, place it into the electrophoresis tank and fill the tank with TAE until the gel is just submerged. At this point, check to see how long the digests have been running. When it has been 1 full hour, proceed.
8. Remove restriction digests from the water bath and add 6X dye to all of them and to an appropriate amount of DNA ladder.
9. Apply the samples carefully to the agarose gel wells. Record the order of the samples and their respective wells.
10. Add 15 μL of ethidium bromide to the positive pole (bottom of tank).
11. Electrophorese the samples at 100 V (constant voltage) for 30 minutes or so, just enough to separate the fragments.
12. Remove the gel carefully from the tank and take it to the gel imaging area along with two razor blades or plastic knives and an ultraviolet (UV)-protective face mask (also wear goggles).
13. Take a photo of your gel and print a copy for each member of your team.

6.1.3.4 Interpretation of Results

1. Observe the gel and classify the results into (a) good, (b) bad, and (c) inconclusive.
2. Throw the bad ones away—you cannot use them. Save the good ones in the –20°C freezer for use in the mammalian cell transfection section.

6.1.4 Expected Results and Pitfalls

The ideal goal is 10 perfect minipreps where your insert can be clearly visualized and where there is an obvious distinction between the new product and the vector digested with the same enzymes. In reality, several other things are likely to happen in addition to, or instead of, the right product. There may be products in the ligation that do not digest or that digest partially, leading to multiple confusing bands of uncut supercoiled plasmid and extracted pieces. There may be size ambiguity between the original vector and the final product. Last, but not least, the new product may recombine in the host cells, "spitting out" some of the plasmid and leading to a final piece smaller than anything expected. This occurs relatively frequently, especially if the plasmid is large or has similar sequences in two different regions. Usually, the best thing to do is simply try again. But, if it continues to happen, special strains of *Escherichia coli* called rec$^-$ can be used to minimize recombination. Figure 6.1 shows two sample gels with desirable and undesirable products.

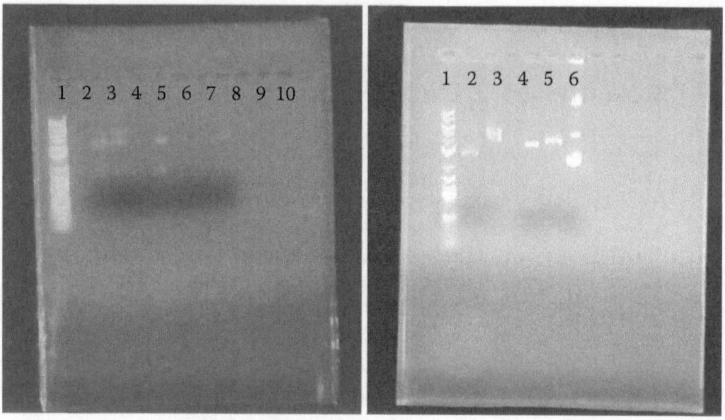

(a) (b)

Figure 6.1 Examples of gels from the ligation performed in this laboratory: (a) A successful ligation. Well 4 shows the desired product. Note, however, that many of the wells are empty (there were 10 minipreps total). Wells 1 and 2 have unexplained bands; they could be partially digested sample, or they could be unwanted products, so it is best to avoid them. Well 7 may have the correct product, but there is too little to be able to tell for sure. (b) Another successful ligation. Here, there were six minipreps only. Well 4 shows the desired product. Well 5 is the original vector (note the subtle size differences between both the vector and the insert). Wells 1 and 6 are just weird and may be uncut, supercoiled plasmid of unknown composition. Throw these away. Well 2 appears to be partially digested original vector—note the two bands in the vector region. It might be something else, too, so throw it out. Well 3 is empty.

Background Reading

Nadeau, J. *Introduction to Experimental Biophysics: Biological Methods for Physical Scientists*. Boca Raton, FL: Taylor & Francis; 2012, Sections 2.5, 2.6

Questions

1. Did your ligation work? If not, what products did you obtain? What would you do differently next time?
2. Estimate the difference in efficiency between electroporation of a ligation and of a supercoiled plasmid as in Experiment 1.
3. If you had some ligations that worked and some that did not, check back to the sketch of your plates in Experiment 5B. Is there any relationship between the size, placement, or appearance of the colonies and the success or failure of the ligation?
4. Why do we have to digest plasmids with restriction enzymes to compare them for size? Why do we not just run them undigested?

7 Use of Software to Design Primers for Cloning Using the Polymerase Chain Reaction (Computer Lab)

7.1 Experiment 7: Use of Software to Design Primers for Cloning Using the Polymerase Chain Reaction (Computer Lab)

7.1.1 Introduction

This lab is a computer exercise designed to show how to design primers for polymerase chain reaction (PCR) cloning. At the end of the exercise, primers will be ordered for Experiment 8.

7.1.2 Materials

A computer with an Internet connection and cloning software is needed.

7.1.3 Experimental Procedures

7.1.3.1 Come Up with a New Cloning Plan

1. Begin with the plasmid sequences you uploaded in Experiment 3.
2. Design a new cloning strategy to move one piece of a plasmid into another plasmid, paying no attention to matching restriction sites. This may be essentially the same as in Experiment 3, but use a different restriction enzyme or enzymes to cut the vector.
3. Pick a site to insert the selected fragment into the new vector.
4. Then, decide what the fragment should look like to insert into that site. Pay attention to start and stop codons, reading frames, and other subtleties.

7.1.3.2 Design Primers

1. Check whether the lab-provided software, or the one you downloaded in Experiment 3, has a primer design tool.
2. If not, look for Primer3 (http://primer3.sourceforge.net/webif.php). It can be downloaded or used as a web interface.
3. Design primers.

7.1.3.3 Primer Subtleties

Check the following and note them:

1. Length of primers (both the part binding to the original plasmid and any restriction "tail").
2. Melting temperatures.
3. Possibility of primer dimer.
4. Do you see any potential problems with these primers? If so, try designing them again.
5. When ready, give the sequences to the teaching assistant for ordering.

7.1.4 Expected Results and Pitfalls

The green fluorescent protein (GFP) family is actually tricky to amplify by PCR. Primer dimer is common. Remember that it is OK to amplify sequences outside your target gene as long as they do not cause problems with reading frame or start/stop or lead to expression of many unneeded amino acids (a few are OK). Check your sequences carefully before finalizing them.

Background Reading

Nadeau, J. *Introduction to Experimental Biophysics: Biological Methods for Physical Scientists.* Boca Raton, FL: Taylor & Francis; 2012, Section 2.7.

Dieffenbach, C. W., T. M. Lowe, and G. S. Dveksler. General concepts for PCR primer design. *PCR Methods Appl* 3:S30–S37 (1993).

Zhou, M. Y. and C. E. Gomez-Sanchez. Universal TA cloning. *Curr Issues Mol Biol* 2:1–7 (2000).

Question

1. These are primers chosen by previous students in the course. Run them through your software and determine whether you think they will work.

   Team 1 primers

   5′ cacc ctg gtt tag tga acc tc

   5′ gag cac aca ctt gca gct ca

   Team 2 primers

   5′ cacc ttt agt gaa ccg tca g

   5′ gag cac aca ctt gca gct c

   Team 3 primers

   5′ cacc tat ata agc aga gct gg

   5′ gag cac aca ctt gca gct

   Team 4 primers

   5′ cacc tag cgc tac cgg t

   5′ gag cac aca ctt gca gct

8 Running the Polymerase Chain Reaction

8.1 Experiment 8A: Running the Polymerase Chain Reaction

8.1.1 Introduction

The objective of today's experiment is to use the polymerase chain reaction (PCR) to generate the *EGFP-f* gene in a sufficient amount to be able to visualize it on a gel, cut out the band, and use it for cloning. Note that the end result of this experiment is not a circular plasmid, as with our previous preps, but a linear piece of DNA with its ends determined by the primers you chose. That means that it cannot be further amplified (it has no promoter or antibiotic resistance gene), so the amount you generate by PCR is all you will have. The resulting fragment will be purified and then sent to a selected facility for direct sequencing without the need to clone further. This is a recent innovation in PCR that saves a lot of cloning.

8.1.2 Materials

1. Primers as ordered from on-site or remote molecular biology supplier
2. Template DNA (from Experiment 2 or instructor's stock)
3. PCR machine
4. PCR kit including deoxynucleotide triphosphates (dNTPs) and polymerase
5. P2 pipettor
6. PCR tubes
7. Materials and supplies for gel electrophoresis

8.1.3 Experimental Procedures

8.1.3.1 Prepare the PCR Reaction

1. Clean your benchtop and gloves carefully with 70% ethanol (EtOH) or (optionally) use the clean bench provided by the instructor. You do not want to amplify the DNA from the turkey sandwich you had for lunch.
2. Get an ice bucket.
3. Remove all of the PCR reagents from the freezer; let them thaw except for the polymerase, which you keep on ice.
4. Prepare a PCR reaction containing the following (in order of addition) (make one reaction for *EGFP-f* and one for the control):

H_2O (to 49.5 µL)
PCR buffer with $MgCl_2$: 5 mL
dNTP mix: 200 mM each at final concentration
DNA template: 1 ng total amount
Primers: 500 pM final concentration each

5. Do **not** add polymerase yet. Wait until everyone is ready. All the students will bring their reactions over to the PCR machine and place them into the holder. The machine will be turned on and heated to 94°C before the polymerase is added; this is called a *hot start* and improves specificity.
6. When the machine is hot, add 0.5 µL of polymerase to each reaction using a P2 pipettor. Close the PCR machine and let it run.

8.1.3.2 Prepare a Gel

Prepare an agarose gel. Each student team needs one large well to purify the PCR product.

8.1.3.3 Purify the Product on a Gel

1. Add 6X loading dye to the entire PCR reaction and place it into a well. Electrophorese for approximately 30 minutes.
2. Take razor blades into the gel room and excise and gel purify your band. Estimate its concentration.
3. Give the DNA to your teaching assistant to be sent out for direct sequencing.

8.1.4 Expected Results and Pitfalls

It is common not to obtain products or to obtain an unwanted product (e.g., primer dimer) on the first round of a PCR experiment. It is also common to obtain insufficient product. If something goes wrong, the key is to figure out why and to explain how you would try again next time. It is also common for direct sequencing to fail, even if your PCR reaction worked. If you obtained multiple bands on your PCR or very little product, it is possible that direct sequencing will not work.

8.2 Experiment 8B: Analysis of Sequencing Results (Computer Lab)

8.2.1 Introduction

The aim of this exercise is to learn to use and interpret sequencing results. The sequencing results will come back to you from the sequencing lab as a string of bases. Unknown bases will be indicated by Ns. The quality of your sequencing data is indicated by the number of Ns (the fewer, the better) and the length of the read.

8.2.2 Materials

1. Sequencing results
2. Computer with cloning software

8.2.3 Experimental Procedures

1. Input the sequence data into the software of your choice. Be careful of how Ns are handled by the selected software.

2. Check the expected sequence against your plasmid map.
3. Check the reported sequence against the published sequence of the gene.

8.2.4 Expected Results and Pitfalls

If you obtained one clear PCR product, you are likely to obtain decent sequencing results. However, the length of the read and the number of Ns will vary depending on sample purity and other factors.

Background Reading

Nadeau, J. *Introduction to Experimental Biophysics: Biological Methods for Physical Scientists*. Boca Raton, FL: Taylor & Francis; 2012, Section 2.7.

Dieffenbach, C. W., T. M. Lowe, and G. S. Dveksler. General concepts for PCR primer design. *PCR Methods Appl* 3:S30–S37 (1993).

Zhou, M. Y. and C. E. Gomez-Sanchez. Universal TA cloning. *Curr Issues Mol Biol* 2:1–7 (2000).

Information about Direct Sequencing

Direct sequencing of PCR products. DNA Sequencing Core. n.d. http://seqcore.brcf.med.umich.edu/doc/dnaseq/pcr.html.

Dorit, R. L., O. Ohara, C. B. Hwang, J. B. Kim, and S. Blackshaw. Direct DNA sequencing of PCR products. *Curr Protoc Mol Biol* Chapter 15, Unit 15 12 (2001).

Wilton, S. Direct sequencing of PCR products. *eLS*. doi: 10.1038/npg.els.0003769. http://www.els.net.

Questions

1. Explain the parameters chosen for the thermal cycling (extension time, etc.). If your target gene were 2 kb long, how would these parameters change?
2. Did you obtain only one band in your PCR reaction? If not, what do you think happened? If you had the experiment to do over, what would you change?
3. How many base pairs of sequence did you obtain from the sequencing lab?
4. Is your sequencing consistent with the expected sequence?
5. Would your sequence results be sufficient to identify the plasmid on a gene search?
6. Why is it critically important to purify the PCR product before sending for direct sequencing?

SECTION TWO

Gene Expression

9 Protein Expression and Purification

9.1 Experiment 9: Protein Expression and Purification

9.1.1 Introduction

The purpose of this experiment is to express a large amount of His-tagged green fluorescent protein (GFP) in an appropriate *Escherichia coli* cell line (BL-21), then extract the expressed GFP protein from the BL-21 cells. In the interests of time, the teaching assistants will prepare the cultures containing the expressed protein; the details are given in this manual for your interest. In class, you will purify the protein using a commercial purification kit; students should skip to Section 9.1.3.3. Because GFP is a soluble protein, we can follow the "native" purification instructions in the kit without too much trouble. We will save aliquots every step of the way for sodium dodecyl sulfate polyacrylamide gel electrophoresis (SDS-PAGE) analysis, so if something goes wrong we will be able to find out where it happened in the next laboratory period.

9.1.2 Materials

1. Chemically competent BL-21 cells
2. Plasmid containing His-tagged fluorescent protein
3. Selective plates and media
4. Absorbance reader
5. Probe sonicator
6. Equilibration/wash buffer (5X)
7. Sterile H_2O
8. Ice bucket and ice
9. Platform shaker
10. Centrifuges
11. Binding resin (one bottle shared among the whole class)
12. A 2-mL purification column
13. Elution buffer
14. Size-exclusion column (prepoured), equilibration and elution buffers, and pump; concentrating filters as needed (optional)

9.1.3 Experimental Procedures

9.1.3.1 Transfect BL-21 Cells with Plasmid of Choice (Done by Teaching Assistant)

1. Transform chemically competent BL-21 cells as in Experiment 1.
2. Plate the cells on a selective plate.
3. Choose a single colony and seed it into a 5-mL volume.

4. After 8–12 hours, seed into a 1-L volume for each student team.
5. Let grow overnight.

9.1.3.2 Induce Protein Expression (Done by Teaching Assistant)

1. Use the absorbance reader to check the optical density at 600 nm (OD600) of your large volume (1-L) culture.
2. When it reaches 0.6–0.8, remove a 1-mL volume as an uninduced control and place on ice.
3. Add 1 mM isopropyl β-D-1-thiogalactopyranoside (IPTG) to the rest and continue shaking for 5–12 hours.
4. Pellet the cells by centrifugation.

9.1.3.3 Purify the Protein on an Affinity Column

1. Drain all medium from the bacterial pellet.
2. Dilute the equilibration/wash buffer to yield a 1X stock (make 50 mL to start).
3. Resuspend the pellet by vortexing in 2 mL of cold 1X equilibration/wash buffer per 25 mL of original culture.
4. Lyse the bacteria by adding lysozyme to 0.75 mg/mL or by three rounds of freeze/thaw in dry ice–ethanol/chilled H_2O.
5. Incubate at room temperature 20–30 minutes (for lysozyme).

6. Meanwhile, get the resin ready. We will use a 2-mL purification column and 2 mL of resin per sample (which corresponds to a 1-mL bed volume). Resuspend the resin, then pipette 2 mL into a sterile 15-mL tube.
7. Centrifuge at 700*g* to pellet the resin, then discard the supernatant.
8. Add 10 mL of 1X equilibration/wash buffer and mix briefly.
9. Centrifuge at 700*g* for 2 minutes, then discard the supernatant.
10. Repeat Steps 8 and 9.
11. After your lysozyme step is finished, sonicate your pellet three times for 10 seconds each with a 30-second pause on ice between each burst.
12. Centrifuge at 10,000–12,000*g* for 20 minutes at 4°C. You can do this in the desktop centrifuge by dividing your sample into 1.5-mL tubes.
13. Carefully transfer the supernatant to a clean tube. This is your **clarified sample**. Reserve a small amount for SDS-PAGE analysis (keep the tube on ice) and add the rest to the tube containing the clean resin.
14. Place on a platform shaker for 20 minutes at room temperature to allow the His-tagged protein to bind the resin.
15. Centrifuge at 700*g* for 5 minutes and remove as much supernatant as possible without disturbing the resin.

16. Wash the resin with 10 mL equilibration/wash buffer. Let it shake for 10 minutes on the platform shaker for thorough washing.
17. Centrifuge at 700*g* for 5 minutes and remove as much supernatant as possible without disturbing the resin.
18. Repeat Steps 16 and 17.
19. Add 1 mL of equilibration/wash buffer to the resin and resuspend by vortexing.
20. Transfer the resin to a 2-mL column with a cap on the end and let it settle out of suspension.
21. Remove the cap and allow the buffer to drain.
22. Wash with 5 mL equilibration/wash buffer.
23. Elute with 5 mL elution buffer in 500-μL increments in clean Eppendorf tubes.
24. Keep the samples at 4°C for further analysis in the next lab period. You might want to peek at them with the ultraviolet (UV) wand to see if they are green.

9.1.3.4 Additional Purification (Optional)

This step of additional purification is necessary if the protein is to be crystallized. Most commonly, a size-exclusion column will be used for further purification. See Chapter 11 and *Introduction to Experimental Biophysics*, Chapter 5, for more on preparing proteins for crystallization.

1. Obtain a prepoured size-exclusion column from the teaching assistant.
2. The resin volume should be 30–100X the volume of the sample. If not, concentrate the sample using a centrifugal ultrafiltration device.
3. Wash the column with at least three volumes of equilibration buffer using a flow rate of 1–1.5 mL/min.
4. Load the sample.
5. Wash with equilibration buffer. As the green color begins to elute, collect and save 2-mL fractions.
6. Save the eluted fractions at 4°C until the next lab period.

9.1.4 Expected Results and Pitfalls

The most challenging aspect of protein binding and elution is ensuring that the protein binds and then choosing the eluted fractions that have the most protein. Because GFP is very green, this will not be an issue in this experiment. You will be able to appreciate which eluate fractions contain the most protein and combine them or throw them out accordingly. It is important to appreciate how difficult this might be for a protein that is not brightly colored. Save different fractions—the results might surprise you in the next section.

Some things can go terribly wrong in this experiment. Protein may not bind to the column, or it may not elute. Contact the teaching assistant if this happens. There may be problems with the resin or buffer.

Background Reading

Nadeau, J. *Introduction to Experimental Biophysics: Biological Methods for Physical Scientists.* Boca Raton, FL: Taylor & Francis; 2012, Chapter 4.

Questions

1. Explain the mechanism of His tag binding to affinity resin. What is the affinity used in this experiment?
2. What is the composition of the different buffers? Why?
3. Are there different types of His tag? If so, do they differ in properties/affinity, and if so, how?
4. Can affinity resin be reused? If so, how?
5. Name and discuss at least one other type of affinity resin besides the one we used.
6. Discuss the types of impurities that might be found in your product.
7. Discuss reasons why protein may not bind or may not elute and what you would do in these cases.
8. Under what circumstances would you use "denaturing" rather than "native" purification conditions? Define each and explain.
9. How does size-exclusion chromatography work? Why are we doing it? (Optional)

10 Characterization of Purified Protein

10.1 Experiment 10: Protein Expression and Purification

10.1.1 Introduction

The purpose of this experiment is to characterize the protein purified in the preceding session by polyacrylamide gel electrophoresis (PAGE) and absorbance/emission spectroscopy. The first step is PAGE, which separates proteins on a polyacrylamide gel, similar to the way agarose gels separate DNA. This will tell you whether your purified protein is consistent in molecular mass with green fluorescent protein (GFP) but tells little about its function. Fortunately, GFP's function is easy to measure. It is fluorescent and absorptive in the visible, so instead of having to analyze it by absorbance at 280 nm (which is complicated by the imidazole in the buffer), it can be quantified using its characteristic absorbance and emission peaks. The combination of techniques will enable you to confirm that you have purified functional GFP from your bacterial pellet.

10.1.2 Materials

1. Purified protein samples from Experiment 9 (including uninduced sample)
2. Sodium dodecyl sulfate (SDS)–PAGE apparatus
3. Protein molecular weight marker (ladder)
4. Acrylamide solution
5. Tris-HCl: 1.5 M, pH 8.8
6. Tris-HCl: 0.5 M, pH 6.8
7. 10% SDS
8. Sample buffer containing bromophenol blue and β-mercaptoethanol
9. 5X running buffer (Tris-glycine, pH 8.3)
10. Coomassie stain and destainer (methanol-40%, acetic acid-10%, distilled water [dH_2O]-50%)
11. Fluorescence plate reader and black plate
12. Absorbance plate reader and clear (quartz) plate

10.1.3 Experimental Procedures

10.1.3.1 SDS-PAGE

1. Wash the gel casting glass plates with H_2O, dH_2O, and then with acetone; wipe dry (be careful; they are fragile).
2. Put the plates into the apparatus and clamp shut. Make sure it is not leaking by adding H_2O.
3. Prepare the resolving gel (7.5% acrylamide). This gel should fill about 75% of the apparatus. In a small beaker, mix the following:

- 2.42 mL dH_2O
- 1.25 mL 1.5 M Tris (pH 8.8)
- 50 mL 10% SDS
- 1.25 mL 30% acrylamide
- 25 mL ammonium persulfate
- 2.5 ml TEMED (tetramethylethylenediamine) (the polymerization reaction starts the instant you add this, so pour into plates right before casting)

4. Pour into the plates, add a layer of ethanol, and wait until the gel solidifies (15–20 minutes).
5. Carefully pour off the layer of ethanol.
6. Prepare the stacking gel in a clean beaker:
   - 3.05 mL dH_2O
   - 1.25 mL 0.5 M Tris (pH 8.8)
   - 50 mL 10% SDS
   - 0.65 mL 30% acrylamide
   - 25 mL ammonium persulfate
   - 5 ml TEMED
7. Pour this onto the gel and add the comb immediately; let solidify for 15 minutes.
8. Meanwhile, prepare the protein samples, including the ladder. Dilute the samples 1:4 with sample buffer and heat at 95°C on a hot plate for 4 minutes. Note: It will smell bad.
9. Remove the cast gel and place it gel-side in into the electrophoresis box.

10. Fill the box with 1X running buffer and run for 30–45 minutes (depending on the dye front, it should reach the bottom) at 200 V.
11. Remove the gel carefully from between the plates using a scoop (provided) and into a small box. Put in Coomassie brilliant blue stain and let sit for 30 minutes.
12. Carefully decant the Coomassie brilliant blue back in the bottle (do not throw out; it is reusable) and add about 100 mL destaining solution.
13. Decant the destainer into a separate bottle (as this is also reused, but not as it is; we need to treat it with activated charcoal).
14. Add 100 mL of destainer again; repeat Step 13.
15. Repeat this until the gel is clear and you see the protein.
16. Scan the gel (with the help of your teaching assistant).

10.1.3.2 Spectroscopy

1. Pipette your samples into the wells of the quartz plate and place in the absorbance reader.
2. Transfer them into a black plate and put them in the fluorescence plate reader. Excite at appropriate wavelengths based on what you saw in the previous step (absorbance measurement).
3. Export the data in an appropriate format for analysis.

10.1.4 Expected Results and Pitfalls

Quite a few things can go wrong during SDS-PAGE. Apart from the pitfalls of gel preparation (solidifying too quickly, tearing, etc.), there are a few other things to keep in mind. The protein heating step may cause some proteins to polymerize and aggregate. If this happens, heat at a lower temperature for a longer time (45°C–55°C for 60 minutes). The orientation of the gel can become confusing once the plates are removed; you may snip a corner off the gel to define its orientation. Coomassie blue is messy, and the staining and destaining often work too well or not enough, respectively. Some practice is required to obtain a good signal without too much dye.

For spectroscopy, it is important to note that higher concentrations are usually needed for good absorbance spectra as compared to what is needed for emission. Outside a certain range, the absorbance will be nonlinear with concentration. This range varies from protein to protein and in fluorescent samples may be affected by protein-protein interactions.

Background Reading

Nadeau, J. *Introduction to Experimental Biophysics: Biological Methods for Physical Scientists*. Boca Raton, FL: Taylor & Francis; 2012, Sections 4.8, 12.1–12.4.

Questions

1. What size is your protein relative to the molecular weight standard?
2. Draw a curve of absorbance versus concentration for your protein. What is the extinction coefficient?
3. Plot the emission spectrum of your protein. Is fluorescence emission intensity linear with concentration?

11 Protein Crystallization

11.1 Experiment 11: Protein Crystallization

11.1.1 Introduction

If we have the time and patience, we would like to try to crystallize the green fluorescent protein (GFP) purified in this laboratory. However, GFP is a rather tricky protein to crystallize, and the protein we have prepared is probably not pure enough to yield satisfying crystals. We will thus use lysozyme as a demonstration protein because it gives good-quality crystals rapidly (<24 hours as opposed to weeks for most proteins). An advanced option is to try with GFP as well; students may take plates home after the course ends if desired.

Crystallization is an important step in determining the three-dimensional structure of proteins. The goal of crystallization is to produce a well-ordered crystal that is lacking in contaminants and large enough to provide an x-ray diffraction pattern. The general process by which crystallization occurs is the same for all proteins, although the ideal conditions vary widely from protein to protein. This process is called vapor diffusion: A droplet containing protein, buffer, and precipitant is allowed to equilibrate with a larger reservoir containing similar ingredients in higher concentrations. Initially, the droplet of protein solution contains an insufficient concentration of precipitant for crystallization, but as water vaporizes from the drop and transfers to the reservoir, the precipitant concentration increases to an optimal level. The system is airtight, and the protein solution either hangs from the ceiling of the chamber (hanging drop) or sits in a pedestal (sitting drop). We will be using crystallization wells designed for the sitting drop method.

11.1.2 Materials

1. Crystallization wells and sealing tape
2. Sodium acetate (1 M solution)
3. Sodium chloride (powder)
4. Lysozyme
5. Polyethylene glycol 300 (PEG 300), citrate, isopropanol, bromophenol blue
6. Purified fluorescent protein from Experiment 9 (optional)
7. Reservoir solutions or ingredients for GFP crystallization (optional)

11.1.3 Experimental Procedures

11.1.3.1 Prepare Reservoir Solutions

Control solution

8% w/v NaCl (0.8 g to make 10 mL)

1 mL of 1 M NaOAc

H_2O to 10 mL

Test solution 1

40% w/v PEG 300

100 mM acetate

200 mM NaCl

Test solution 2

35% v/v isopropanol

100 mM acetate

11.1.3.2 Prepare Lysozyme for Each Student Team

Each team will receive a vial of lysozyme and some solubilization buffer (the buffer is 0.02 M sodium acetate, pH 4.6). Dissolve your lysozyme in buffer as follows:

Team 1: Add 1 mL of solubilization buffer (lysozyme concentration: 20 mg/mL)

Team 2: Add 0.8 mL of solubilization buffer (lysozyme concentration: 25 mg/mL)

Team 3: Add 0.5 mL of solubilization buffer (lysozyme concentration: 40 mg/mL)

Team 4: Add 0.4 mL of solubilization buffer (lysozyme concentration: 50 mg/mL)

Team 5: Add 0.2 mL of solubilization buffer (lysozyme concentration: 100 mg/mL)

More than 5 teams: Repeat from Team 1

11.1.3.3 Prepare Crystallization Wells with Different Conditions

The following is the procedure for each condition:

1. Select crystallization well.
2. Load reservoir with 600 μL reservoir solution.
3. Load pedestal with 7.5 μL reservoir solution (taken from reservoir).
4. Load pedestal with 7.5 μL protein solution.
5. Cover well with sealing tape.

Try the following conditions:

1. Normal (control solution is the reservoir solution).
2. Using test solutions 1 and 2 as the reservoir.
3. Boiled. Take 50 μL of your protein solution and heat it in a hot plate for 3 minutes at 95°C. Let cool, then load.
4. Freeze/thaw. Transfer five 50-μL aliquots of protein solution to five microfuge tubes. Label them 1–5. Freeze all five tubes; remove from freezer and allow to thaw.

Return tubes 2–5 to freezer and repeat until each tube has been frozen and thawed as many times as its number.

5. Foreign agent. Add 0.5 μL of 1% w/v bromophenol blue to protein solution.
6. Physical torture. Shake your protein solution around really hard, filter it, or centrifuge it. Note what you did.
7. Let the plates sit at least overnight before inspecting.

11.1.3.4 Crystallization of Ultrapure Green Fluorescent Protein (Optional)

The crystallization of ultrapure GFP is a highly advanced option. The experimental procedures are as discussed previously, but the composition of the buffer and reservoir solutions must be chosen with great care. Some references are included at the end of the chapter to help guide you. At least four 24-well plates should be prepared using different reservoir solutions, protein concentrations, and incubation temperatures. See *Introduction to Experimental Biophysics,* Chapter 5, for more details.

11.2 Experiment 11B: Inspection of Crystals with Polarization Microscopy

11.2.1 Introduction

Light microscopy is used to identify protein crystals. The use of polarizers is especially helpful for distinguishing crystals from other artifacts. The variety of conditions used in the crystallization experiment will result in crystals of varying size and quality.

11.2.2 Materials

1. Plates from Experiment 11A
2. Polarizing microscope

11.2.3 Experimental Procedures

1. Examine the plates under the microscope.
2. Rotate the polarizer to observe the difference it makes.
3. Take photomicrographs of the different crystallization conditions.

11.2.4 Expected Results and Pitfalls

Lysozyme is one of the easiest proteins to crystallize. In most of the conditions tested, you should obtain three-dimensional crystals that are readily indentified under the microscope. Some of the conditions, such as the physical shock, may prevent crystallization entirely; this will serve as an interesting control. Although this is an "easy" experiment, things can go wrong. The most common errors seen are

- Unlabeled plates. Label both the lid and the plate itself with the sample name, number, or other designator. This way, if the lid is lost or mixed up with that from another plate, it is still clear which sample is which.
- Precipitation caused by sudden addition of concentrated protein into precipitant. Pipette crystallization

solution into protein solution rather than the other way around.

- Contamination. Always change pipette tips between samples and wear gloves and a lab coat. Remove any hairy/fuzzy sweaters.
- Harsh treatment of protein: heating by exposure to body heat or excessive room heat, dropping the vial, or contaminants/dust from the air falling in. Treat the sample gently and keep it on ice when not in use.
- Running out of reagents before the plate is fully prepared. Make sure you prepare enough protein and reservoir solutions. This is especially important when the solutions are viscous because a good deal is lost during pipetting in this case.
- Poor sealing of the crystallization plate. Obtaining a good seal is key. The tape provided is just packing tape, but it must be made to seal onto the plate without bubbles or holes. Practice pressing and rolling it down with your gloved fingers or implements. Find what works best for you.

Background Reading

Nadeau, J. *Introduction to Experimental Biophysics: Biological Methods for Physical Scientists*. Boca Raton, FL: Taylor & Francis; 2012, Chapter 5.

For Advanced Option

For crystallization of fluorescent proteins (some examples of journal articles containing recipes):

Battistutta, R., A. Negro, and G. Zanotti. Crystal structure and refolding properties of the mutant F99S/M153T/V163A of the green fluorescent protein. *Proteins* 41:429–437 (2000).

Griesbeck, O., G. S. Baird, R. E. Campbell, D. A. Zacharias, and R. Y. Tsien. Reducing the environmental sensitivity of yellow fluorescent protein. Mechanism and applications. *J Biol Chem* 276:29188–29194 (2001).

Ormo, M., A. B. Cubitt, K. Kallio, et al. Crystal structure of the *Aequorea victoria* green fluorescent protein. *Science* 273:1392–1395 (1996).

Wilmann, P. G., K. Turcic, J. M. Battad, et al. The 1.7 A crystal structure of Dronpa: a photoswitchable green fluorescent protein. *J Mol Biol* 364:213–224 (2006).

Questions

1. Define vapor diffusion, interface diffusion, microbatch, and dialysis.
2. Define precipitant and mother liquor.
3. Paste images of your lysozyme crystals into your notebook. How did the different conditions tested affect crystallization?
4. What makes crystals grow to different shapes and sizes?
5. How does polarization microscopy help to identify crystals? Can it be used to distinguish protein crystals from salt?

6. What are some other ways to distinguish protein crystals from unwanted contaminants?
7. How would you go about beginning a crystallization experiment for a protein such as GFP? Where would you begin, and how would you vary the parameters?
8. "Score" your GFP crystals according to the guide in *Introduction to Experimental Biophysics,* Chapter 5 (advanced option only).

SECTION THREE

Fluorescent Reporter Genes in Mammalian Cells

12 Transfecting Mammalian Cells with Custom-Designed Constructs

12.1 Experiment 12: Transfecting Mammalian Cells with Custom-Designed Constructs

12.1.1 Introduction

After the cloning experiments of Chapters 5 and 6, you should have a new and interesting construct in a mammalian expression vector. Depending on what you made, you will be able to screen it in several ways: by fluorescence microscopy if it is fluorescent, by electrophysiology if it is an ion channel, and so on. We will discuss the best way to do this, but the first step is to get it into cells.

Thousands of "immortalized" mammalian cell lines exist. These cells are derived from cancers or contain oncogenes that allow them to reproduce essentially indefinitely. They are purchased as a single frozen aliquot (the primary supplier is the American Type Culture Collection or ATCC) and then propagated by "passaging": When the cells become too crowded or confluent, they are removed from the dish and split into several dishes. If the cells are adherent, meaning they stick to the bottom of the dish, the enzyme trypsin is used to loosen them and facilitate the passaging. The concentration of trypsin used and the length of time it needs to be used on the cells varies tremendously among cell lines; some stick so that they are hard to remove, and others come up easily. The example cells we will use in this lab tend to release easily.

The goal of this experiment is to passage a mammalian cell line into several small dishes or wells and then to transfect your new construct either alone or with other plasmids to make it visible, inducible, and so on. The key to working with mammalian cells is *sterile technique*. Lack of sterility causes the cell lines to become infected with bacteria or fungi from the air and can infect the entire incubator and be difficult to eradicate. Even baker's yeast can infect a cell line.

The general rules are as follows:

- Your gloves are dirty. Anything you touch with a gloved hand is dirty and must not touch medium or cells.
- A pipette that touches anything but medium or cells is dirty.
- Any tool that passes out of the laminar flow hood is dirty. Never pass items in and out of the hood.
- One person at a time should be in the laminar hood or the flow will be disrupted.

- Sterilize gloves, bench surfaces, and so on with 70% ethanol.
- Any dish with spilled medium is dirty and must be discarded.

Please warn the instructor if you see any signs of contamination in the cells. This includes cloudy medium, medium turning yellow, or visible colonies of mold.

After passaging the cells and allowing them to reattach, we will introduce a plasmid or plasmids into the cells using a liposomal transfection method. What is fortunate about this method is that two (or even more) plasmids can be mixed together in equal ratios and will "cotransfect" into cells: That is, a cell that receives one plasmid must receive the other. This opens many possibilities for creativity in these experiments.

The default experiment in this lab is to explore tetracycline-inducible expression using the pTet-On system (revisit Experiment 1A, Figure 1.1, and the mystery plasmids for details). To obtain expression that turns on in the presence of tetracycline, the plasmid pTet-On must be cotransfected along with the plasmid pTre that contains the gene of interest. The original pTre plasmid provided did not contain any gene of interest. Your new construct may contain a red fluorescent protein, a green fluorescent protein, or something else (if you provided something else). The original mystery plasmids encoding red and green fluorescent proteins may be used as positive controls for your transfection technique, as they should express at high levels without any cotransfection or induction necessary. Thus, several different dishes are needed for each team to explore the different possible combinations using these plasmids.

12.1.2 Materials

1. Flask of a mammalian cell line (probably HEK293 cells)
2. Plastic- or glass-bottomed culture dishes or (recommended) 12-well plate
3. Sterile plasticware for tissue culture
4. Tissue culture hood with biohazard disposal
5. Culture medium (with serum) and serum-free medium (Opti-MEM®)
6. Sterile phosphate-buffered saline (PBS)
7. Trypsin
8. Liposome-based transfection reagent (LipofectAMINE [Invitrogen])
9. pTreTight-reporter gene DNA (prepared by ligation or ask instructor if your ligation failed)
10. pTet-On
11. pEF1α-AcGFP-N1 and pDsRed2-Mito (for positive control; refer to Figure 1.2)

12.1.3 Experimental Procedures

12.1.3.1 Passaging Cells for Transfection

1. Obtain a flask of cells. Aspirate the medium from the flask using a Pasteur pipette and vacuum system inside the tissue culture hood.
2. Using a sterile 5- or 10-mL pipette, add 1 to 2 mL PBS, swirl the flask, and suck off the PBS again. This removes traces of serum that may inhibit the action of trypsin.
3. Add 1 mL trypsin solution and gently rock the flask for 30 seconds.
4. Aspirate the trypsin. Close the flask cap and let it sit in the hood for 2 minutes.
5. Fill a 10-mL pipette to 6 mL with fresh growth medium. Squirt into the flask; you should be able to see the cells begin to detach. Pipette up and down about 10 times or until you see a homogeneous cell suspension inside the pipette.
6. Place 0.5 mL (or about 8 drops) of the cell suspension into each dish.
7. Fill the wells with medium (~1.5 mL for a 12-well plate).
8. Put the plates in the 37°C CO_2 incubator and leave at least overnight.

12.1.3.2 Transfecting Cells with New Construct

Everything will be done in a tissue culture (laminar flow) hood. This means do **not** open the liposomes or any other reagent outside the hood or there will be contamination; all your cells will die.

These instructions are based on a 12-well plate with the setup shown in Figure 12.1. The quantity of reagents must be adjusted if other well sizes are used.

1. Collect the plasmids you need and bring them into the tissue culture room. Make sure you know their concentrations.
2. Each well of the 12-well plate requires 0.5 μg of DNA (**total**, regardless of the number of plasmids) and 5 μL of liposomes. Follow Figure 12.1; note that you have the following:

   1 well with no DNA (well 1)

   1 well with positive control (pEF1α-AcGFP-N1 or pDsRed2-Mito; you can combine both if you wish!) **(well 2)**

   1 well with pTet-On only (well 3)

   1 well with pTre-reporter gene only (well 4)

   8 wells with pTre-reporter gene and pTet-On (wells 5–12)

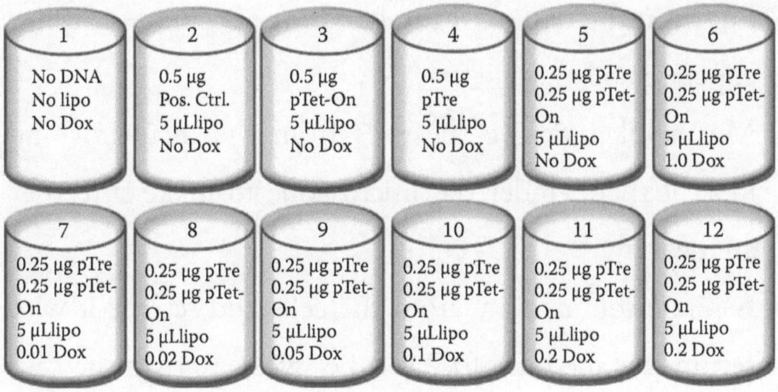

Figure 12.1 Set up of 12-well plate for inducible gene expression experiment.

So, prepare one set of solutions containing the following:

25 µL of Opti-MEM medium only

0.5 µg of positive control plasmid in 25 µL of Opti-MEM

0.5 µg of pTet-On in 25 µL of Opti-MEM

0.5 µg of pTre-reporter in 25 µL of Opti-MEM

2.0 µg pTre-reporter and 2.0 µg pTet-On in 0.2 mL Opti-MEM

Prepare a second solution containing

24 µL of liposomes in 0.3 mL of Opti-MEM

3. Mix the solutions gently, then pipette 27 µL of the liposome solution into each of the first four DNA solutions. Add the rest to the final DNA solution.

4. Let sit in the tissue culture hood for 30 minutes. It may appear cloudy.
5. Meanwhile, remove the cells from the incubator and look at them under the microscope to make sure they are healthy.
6. Aspirate the medium from the cells and replace it with serum-free medium (0.2 mL per well). Return the cells to the incubator.
7. When the liposome solutions are ready, pipette 150 μL of Opti-MEM into each individual reaction (1.2 mL into the eightfold reaction). Pipette gently several times, then add to the cells. Return the cells to the incubator.

12.2 Experiment 12B: Adding Serum and Inducing with Doxycycline

12.2.1 Introduction

Most transfection reagents do not work in the presence of serum. However, the cells will die if left too long without serum. At some point after the liposomes have gotten into the cells (after 5–24 hours), serum needs to be added. The gene will express 24–48 hours after that. Doxycycline may be added 1–12 hours before inspecting the cells. It should not be added more than 12 hours before final examination of the cells because it will degrade. Thus, these steps should probably be performed on subsequent days. The teaching assistant may perform one or both of these steps in the interest of scheduling.

12.2.2 Materials

1. Culture medium containing serum
2. Doxycycline stock solution

12.2.3 Experimental Procedures

1. At 5–24 hours after transfection, aspirate the liposome-containing medium off the cells. Replace with full-serum medium.
2. At 1–12 hours before Experiment 13, add doxycycline to the concentrations given in Figure 12.1 (concentrations are in micrograms per milliliter).
3. Work in the tissue culture hood and return the cells to the incubator promptly.

12.2.4 Expected Results and Pitfalls

The expected outcome of this experiment is a plate of cells glowing perfectly with fluorescent proteins, green and red—all except the controls, which will be healthy but nonfluorescent. This best-case scenario is possible, but only if all reagents are handled with extreme care. Transfection is a complex process: We are taking new genes, inserting them into living cells, and expecting the cells to express these genes (sometimes in an inducible manner). This technique forms the basis of most nonviral approaches to gene therapy, so it is the object of much intensive study, both in vitro and in vivo. Although we are using reasonably straightforward systems—commercial plasmids and liposomes, cell lines optimized for high expression levels—things still can and do go wrong.

The biggest pitfall in mammalian cell work is always contamination. Sterile technique cannot be emphasized more. Even if you manage to avoid gross contamination (and it is gross), it is still possible to have sick cells. Cells that refuse to adhere to the dish, round up, appear grainy, or do not multiply are likely infected. It is also possible to kill cells by rough handling, overtrypsinization, leaving them outside the incubator for too long, forgetting serum, or adding too much liposome solution. The cells are alive—treat them gently.

It is also possible for the cells to be alive and healthy, but for the transfection to have failed. One of the most common reasons for this is miscalculation of DNA or liposome concentration. Poor-quality DNA or faulty pipetting may lead to the same result. All teams in the course should receive the same DNA stock for the positive control. If the expression levels in the positive controls vary widely, then faulty technique is to be suspected.

One more issue here deserves mention: Doxycycline is fluorescent. At the low concentrations we use for induction, this fluorescence is negligible. However, mistakes in dilution of the doxycycline stock can lead to spurious fluorescence.

Background Reading

Nadeau, J. *Introduction to Experimental Biophysics: Biological Methods for Physical Scientists.* Boca Raton, FL: Taylor & Francis; 2012, Sections 3.3–3.5.

Information about the Tet System

Clontech Laboratories. Tet-Off and Tet-On Gene Expression Systems user manual. http://clontech.com/xxclt_searchResults.jsp?searchTerm=tet-on

Questions

1. Why are transfections performed in serum-free medium?
2. What is the principle behind liposome-mediated transfection? Can you think of other compounds or molecules that would be capable of the same process?
3. Why buy commercial transfection reagents if we know they are just liposomes? Could we make "homemade" liposomes? How would you go about this?
4. What are some other ways of getting DNA into cells?
5. Does the DNA get into the nucleus? If not, how does it lead to protein production? Why does this work, and how could it go wrong?
6. Why does contamination make the medium turn yellow?

13 Screening Fluorescent Reporter Gene Expression in Mammalian Cells

13.1 Experiment 13: Screening Fluorescent Reporter Gene Expression in Mammalian Cells

13.1.1 Introduction

The cells containing a fluorescent protein under the control of the tetracycline-inducible promoter should now be expressing different levels of fluorescence depending on the amount of doxycycline added. The purpose of today's lab is to examine the expression levels in two ways: (1) qualitatively, using epifluorescence microscopy with a charge-coupled device (CCD) camera image capture and (2) semiquantitatively using a fluorescence plate reader.

13.1.2 Materials

1. Cells from Experiment 12
2. Sterile phosphate-buffered saline (PBS) and serum-free medium (Opti-MEM®)

3. Fluorescence plate reader
4. Epifluorescence microscope with CCD camera

13.1.3 Experimental Procedures

13.1.3.1 Fluorescence Emission Using the Plate Reader

1. Wash the cells twice with PBS to remove all traces of the pink medium (the phenol red in it is fluorescent); place in Opti-MEM.
2. Place the plate in the plate reader.
3. Read the spectra from all of the wells in the plate. Take both excitation and emission spectra at appropriate wavelengths. If both red and green proteins are present, take spectra for all wells at both the red and green protein excitation wavelengths.
4. Save the data to take home for analysis.

13.1.3.2 Epifluorescence Microscopy

1. Observe the features of the microscope. Note the different excitation and emission filter sets and their properties.
2. Find the cells under low power. Take at least one photo of each of the wells using the appropriate filter set. Save to a compact disk (CD) and bring with you.

13.1.4 Expected Results and Pitfalls

The expected outcomes of this experiment are emission spectra and images of cells containing fluorescent proteins. The biggest pitfall in this experiment is that the plate reader may not be sensitive enough to detect emission from the wells. If this is the case, after microscopy, trypsinize the cells, centrifuge them gently to pellet, and then read the emission in a cuvette.

Fluorescence microscopy is an art, and it will take some practice to be able to take nice images of cells expressing fluorescent proteins. It is always a good idea to take a bright-field image for comparison with the fluorescence image. It is also a good idea to start by looking at the control plate: If it is highly fluorescent under any of the filters, this is background autofluorescence that must be eliminated. Using a neutral density (ND) filter can reduce excitation power enough to cut down autofluorescence without compromising fluorescent protein excitation. A narrower bandpass filter might also be needed; compare different filters as available.

After being outside the incubator for an hour or two, the cells will begin to die. There is not much that can be done about this, except to remove the cells from the incubator only when the microscope is ready.

Background Reading

Nadeau, J. *Introduction to Experimental Biophysics: Biological Methods for Physical Scientists*. Boca Raton, FL: Taylor & Francis; 2012, Chapter 6 and Section 12.4.

Questions

1. Create a plot of excitation and emission spectra for your positive control (well 2) as well as one of the induced wells (e.g., well 12). Do they differ?
2. Create a plot of peak emission versus doxycycline concentration.
3. Do you have a lot of autofluorescence? If so, how could this be reduced?
4. How quantitative do you think the plate reader is? What are its weaknesses?
5. How good is the agreement between the plate reader results and the appearance of the cells taken under fluorescence microscopy? Paste a picture of your fluorescence micrographs in your lab book and comment.
6. Where is the green fluorescent protein or dsRed localized in these cells? Why? How would you prove that this localization occurred as expected?
7. Can you tell which cells are healthy and which ones are sick or dying? Do the sick ones have a different emission spectrum? Does this affect the plate reader results?
8. How do the different fluorescence filters affect the autofluorescence spectrum?

SECTION FOUR

Nanoparticles

14 Synthesis of Quantum Dots

14.1 Experiment 14: Synthesis and Spectral Characterization of Quantum Dots

14.1.1 Introduction

Quantum dots (QDs) are fluorescent semiconductor nanocrystals whose emission peak wavelength is proportional to the size of the crystal. QDs have gained tremendous popularity in recent years for biomedical, alternative energy, and electronics applications. Their popularity was sparked by the development in 1996 of a method of size-selective synthesis: Homogeneous size populations—representing narrow fluorescence emission peaks—could be made directly, without separation steps. A whole range of sizes and colors could be made in a single experiment, thus opening the door to multicolor labeling for all applications. The most common type of QD has a core of cadmium selenide (CdSe) because the band-gap energy of this material makes particle sizes of 2–6 nm span the visible spectrum.

The original CdSe QD synthesis methods used dimethylcadmium as a precursor, which is highly toxic. Less-toxic cadmium precursors can also yield excellent QDs, particularly CdO. Low-temperature, aqueous, "green" methods have also been reported for QD synthesis, but these generally lead to poorer crystal structures and lower fluorescence quantum yields than the organometallic routes. The purpose of this laboratory is to demonstrate high-temperature organometallic synthesis of CdSe QDs and then to distribute the different colors to the student teams for characterization using absorbance and emission spectroscopy or fluorescence microscopy.

14.1.2 Materials

1. QD synthesis setup with Schlenk line (demo)
2. Cadmium and selenium precursors
3. Solvent such as trioctylphosphine oxide (TOPO) for synthesis
4. Solvents for QD precipitation and resuspension (methanol, toluene)
5. Absorbance and emission spectrometers
6. Quantum yield reference standard (e.g., rhodamine) (optional)
7. Fluorescence microscope (optional)

14.1.3 Experimental Procedures

14.1.3.1 QD Synthesis Demo

1. Prepare the selenium precursor in a sealed vial with stirring.

2. Add CdO, degas, and heat to a selected temperature (usually 250°C).
3. Turn off the heat mantle and rapidly inject the selenium precursor.
4. Withdraw 1-mL aliquots at selected time points and quench in cold toluene.
5. Optional: Precipitate and wash several times.
6. Take a photo of the size series under room light and ultraviolet (UV) wand illumination.

14.1.3.2 QD Characterization

1. Each team should receive one QD size/color (double up teams as needed).
2. Measure absorbance and emission.
3. Prepare emission-versus-concentration curves with excitation at 400 nm. Calculate quantum yield relative to a provided standard (optional).

14.1.3.3 Fluorescence Microscopy of QDs (Optional)

1. Dilute the QDs until they are colorless.
2. Place a 5- to 10-μL droplet on a glass microscope slide or coverslip and spin- or air-dry.
3. Place in the fluorescence microscope.
4. Observe properties: appearance under different filter sets; aggregation; fluorescence intermittency (blinking).

14.1.4 Expected Results and Pitfalls

Depending on the initial temperature of the reaction and the speed with which the first sample is withdrawn, the smallest QDs might be blue, turquoise, or green. Obtaining blue is difficult, so do not be surprised if the smallest sample is green. Long cooking times (30–60 minutes) are needed for deep-red QDs; often, these show poorer homogeneity and crystal structure than the smaller QDs. The quantum yields at both size extremes will likely be significantly lower than the "midrange" colors (green, yellow, orange).

It is relatively easy to make QDs—provided that all the necessary apparatus is in place and the safety precautions are followed—but it can be challenging to produce high-quality QDs with high quantum yields and narrow size distributions. The characterizations done in the experiment are intended to give a good idea of the quality of the QDs produced.

When performing spectroscopy, it is important to note that a much higher concentration is needed to obtain a good absorbance spectrum than is needed to take a fluorescence emission spectrum. Make sure that the first exciton peak is visible in the spectrum that you take, then dilute the sample for emission spectra.

To see QD blinking, the sample must be dilute, on the order of 1 nM or even less. A wafer spinner, if available, aids substantially in depositing good individual QDs for blinking studies.

Background Reading

Nadeau, J. *Introduction to Experimental Biophysics: Biological Methods for Physical Scientists.* Boca Raton, FL: Taylor & Francis; 2012, Chapter 8; also see Chapter 6 for quantum yield techniques.

Also see references at the end of Chapter 8 for classic QD synthesis papers.

Questions

1. Why is it critical to synthesize QDs under inert gas?
2. What are the absorbance and emission peaks of your QDs? What is the Stokes shift?
3. What do these spectra say about the mean size and size distribution?
4. Plot the absorbance spectrum and indicate the first exciton peak. Are there other peaks? To what do these correspond?
5. What is the quantum yield of the QD sample? (Optional)
6. Show a micrograph of the QDs. Comment on the color distribution and appearance. Can you tell if you have individual QDs as opposed to aggregates? If so, how? Did the QDs blink? Did the blinking change with time? (Optional)

15 Synthesis of Gold Nanoparticles

15.1 Experiment 15: Synthesis of Gold Nanoparticles

15.1.1 Introduction

Gold nanoparticles have size-dependent optical properties that differ from those of semiconductors but are extremely useful for imaging and sensing applications. Gold nanoparticles are among the easiest nanomaterials to synthesize: Reduction of gold salts by a selected reducing agent yields particles soluble in water or organic solvent as desired. The size of the particles during synthesis is controlled by changing the amount of reducing agent present in the reaction. Different particle sizes appear different colors because of changes in the plasmon resonance frequency caused by damping of electronic oscillations as particles become smaller. The characteristic color of 5-nm gold particles is a striking ruby red. Larger particles appear blue. The purpose of this exercise is to synthesize gold nanoparticles by sodium citrate reduction in aqueous solution. This results in gold nanoparticles about 15 nm in diameter, capped with citrate.

15.1.2 Materials

1. Round-bottom flask
2. Oil bath or hot plate, stir plate, stir bar
3. Hydrogen tetrachloroaurate ($HAuCl_4$) and sodium citrate ($Na_3C_6H_5O_7$)
4. Ultrapure water (18 MΩcm)
5. Ultraviolet-visible (UV-Vis) spectrophotometer
6. Salt solutions (NaCl, KCl)
7. Dark-field microscope or laser confocal microscope (optional)

15.1.3 Experimental Procedures

15.1.3.1 Gold Nanoparticle Synthesis

1. Dissolve 20 mg $HAuCl_4$ in 100 mL ultrapure water in the round-bottom flask.
2. Place into the oil bath and bring to a boil with constant stirring.
3. Dissolve 100 mg of sodium citrate in 2 mL ultrapure water.
4. Add the citrate solution rapidly to the boiling gold solution.
5. Wait until the solution turns red, then allow to cook for 10 more minutes.
6. Remove from heat and allow to cool.

15.1.3.2 Gold Nanoparticle Characterization

1. Dilute the solution appropriately and measure the absorbance spectrum.
2. Add other salts to the solution, such as NaCl or KCl, and observe a change in the spectrum.

15.1.3.3 Dark-Field or Reflectance Microscopy of Gold Nanoparticles (Optional)

1. Deposit a layer of gold nanoparticles on a microscope slide.
2. Observe with dark-field microscopy or confocal microscopy in reflectance mode.
3. Obtain images.

15.1.4 Expected Results and Pitfalls

Good, brisk stirring is necessary to obtain homogeneous-size particles. It is possible to do this experiment without an oil bath by simply placing a flat-bottom flask on a hot plate. However, this will compromise homogeneity, and it is necessary to make sure the solution is not boiled dry. Altering the ratios of citrate to gold can change particle size somewhat; if more than one synthesis per class is performed, this is an interesting avenue to explore.

In general, this is a straightforward experiment that should go well. The particles produced will be a deep, purplish-red, not ruby red.

Background Reading

Nadeau, J. *Introduction to Experimental Biophysics: Biological Methods for Physical Scientists.* Boca Raton, FL: Taylor & Francis; 2012, Chapter 9.

Questions

1. Can you estimate the size of the gold particles from the absorbance spectrum? How about the concentration?
2. What is the best way to determine the concentration of a solution of gold nanoparticles?
3. What is happening when salts such as NaCl or KCl are added to the gold nanoparticle solution? How can this principle form the basis for a biosensor?
4. Show micrographs of dark-field or reflectance microscopy and comment on appearance (optional).

SECTION FIVE
Electrophysiology

16 Electrophysiology: Lipid Bilayer Recording

16.1 Experiment 16: Lipid Bilayer Recording

16.1.1 Introduction

The aim of this exercise is to introduce you to standard techniques of lipid bilayer formation, ion channel insertion, and single-channel recording. You will learn to form a lipid coat across a Teflon® orifice, monitor bilayer formation by capacitance change, and insert a simple pore-forming channel. Standard techniques of single-channel data analysis are introduced.

16.1.2 Materials

1. 1-palmitoyl-2-oleoyl-phosphatidylethanol amine (POPE) and palmitoyl-oleoyl phosphatidylserine (POPS) in chloroform
2. Polished glass rods
3. Decane
4. Hamilton syringe and glass vials

5. A “mystery channel,” which may be alamethicin, gramicidin, or valinomycin (provided as stock solution)
6. Delrin® cup and Teflon chamber
7. Nitrogen gas
8. “Internal” and “external” recording solutions (note composition in your lab book)
9. Silver electrodes
10. 3 M KCl
11. Agar bridges

16.1.3 Experimental Procedures

16.1.3.1 Preparation of Lipids

Note: Work with all chloroform solutions in the chemical hood. Keep lipid vials open for as little time as possible to prevent entry of water vapor.

1. Using a Hamilton syringe, measure 26 μL of POPS into a glass vial. Rinse the syringe with chloroform, then measure in 54 μL of POPE. Swirl to mix.
2. Seal the lipid stock bottles tightly and return them to the freezer.
3. Using a gentle stream of nitrogen gas, evaporate the chloroform from the lipid. Try not to splash; you should be able to see the dried lipid at the bottom of the vial when you are finished.
4. When the lipid is completely dry, measure 100 μL of decane into the vial. Swirl until the lipid is redissolved.

5. Take the lipid into the physiology room along with the glass rods, recording solutions, and ion channel stock solution.

16.1.3.2 Rig Setup and Bilayer Formation

1. Arrange all of the solutions in the physiology room. Share rigs as appropriate.
2. Proceed with rig setup (see Background Reading section): Turn on the amplifier and start the software; attach the silver electrodes and agar bridges; place the Teflon chamber into the Faraday cage; put the Delrin cup into the chamber; fill each side of the chamber with internal and external solutions.
3. Ensure that you have the appropriate open-channel signal on the software (and that you understand what this means).
4. Locate the orifice in the Delrin cup under the dissecting microscope.
5. Quickly dip a glass rod into the lipid-decane mixture (close the cap afterward; oxygen will destroy the lipid). Rub it across the orifice until there is zero current flow.
6. Wait a minute or two to see if a bilayer forms spontaneously. If not, "probe" the orifice with the rod until either (a) you obtain a bilayer or (b) the lipid breaks. If it breaks, try to re-form it.

16.1.3.3 Ion Channel Assembly

1. When a stable bilayer is present, add the ion channel in the recommended concentration.
2. Wait a few minutes to see if channel-opening behavior is observed.
3. If there is nothing, break the bilayer and re-form it.
4. When channel events are seen, record them for at least 5 minutes. Change the recording parameters to observe the effects on the signal.

16.1.4 Expected Results and Pitfalls

The provided ion channels are highly robust and should give nice signals. Forming a bilayer is challenging but, under the conditions specified, should be achievable in the allotted lab time. Be patient.

One of the biggest problems with channels such as valinomycin is that it is hard to get rid of them. If added in excessive concentrations, the ionophore will destroy the bilayers immediately and be nearly impossible to wash out. If this appears to have occurred, take the sample cup and chamber and wash with a dilute detergent solution followed by copious amounts of water. Then, try everything again.

Ionophore solutions can also go bad. If nothing is seen with what should be a sufficient concentration of channel, ask the teaching assistant for a fresh solution.

Background Reading

Nadeau, J. *Introduction to Experimental Biophysics: Biological Methods for Physical Scientists*. Boca Raton, FL: Taylor & Francis; 2012, Chapter 11.

Information about Lipid Bilayers

Molecular Devices. *Axon Guide: Electrophysiology and Biophysics Laboratory Techniques*. 3rd ed. Sunnyvale, CA: Molecular Devices; 2012.

Available for free download from a number of sources. Provides a general introduction to electrophysiology; highly recommended.

Warner Instruments. Planar lipid bilayer workstation. https://www.warneronline.com/product_info.cfm?ID=1094.

Highly recommended.

Other Recommendations

Ashley, R. H. (ed.). *Ion Channels: A Practical Approach*. New York: Oxford University Press; 1995.

Rudy, B. (ed.). *Ion Channels*. Maryland Heights, MO: Elsevier; 1992.

Questions

1. Which ion channel did you have?
2. How does the channel you worked with function? How is it different from large protein channels?
3. Why do we use capacitance to monitor bilayer formation? Is this a reliable method for all orifice sizes? What is the best way to measure capacitance electronically, or does it matter?

4. What determines the recording parameters (acquisition speed, filter rate, etc.)? What happens to the data when you vary these parameters?
5. Did you obtain one channel or more than one? How do you tell the difference?

17 Electrophysiology Alternative: Introduction to Lipid Vesicles

17.1 Experiment 17: Introduction to Lipid Vesicles

17.1.1 Introduction

The principle behind this experiment is to demonstrate how to evaluate the changes in the spectrum of a voltage-sensitive dye using a membrane potential established across the membrane of micron-size lipid vesicles. The membrane potential is established using an ionic balance and a selective ionophore. This type of approach to creating a membrane potential is useful for experiments in which planar lipid bilayers are not ideal, for example, where high-resolution microscopy is needed or the experiment needs to be performed in a small volume. The general approach to the use of liposomes can be used in a wide variety of applications testing pumps or other channels for their ability to move molecules from one side of a membrane to another or testing dyes for ionic sensitivity.

17.1.2 Materials

1. Lipid (1,2-Dioleoyl-sn-glycero-3-phosphocholine [DOPC] or phosphatidylcholine [PC]) in chloroform, Hamilton syringe
2. "Internal K" solution (high KCl) and "external K" solution (low KCl); "internal Ca" solution (high $CaCl_2$) and "external Ca" solution (low $CaCl_2$)
3. Glass vials and cryo vials
4. Nitrogen gas and liquid nitrogen; sonicator
5. Voltage-sensitive dye (recommended: RH421 or di-4-ANEPPS) in stock solution
6. Calcium-sensitive dye (Fluo-3) in stock solution
7. Valinomycin and calcium ionophore in stock solution
8. A 96-well black plate

17.1.3 Experimental Procedures

17.1.3.1 Preparation of Vesicles

Note: Work with all chloroform solutions in the chemical hood. Keep lipid vials open for as little time as possible to prevent entry of water vapor.

1. In the chemical hood, measure 100 μL of lipid into each of two glass vials using the Hamilton syringe. Recap the lipid promptly and return it to the freezer.
2. Dry the lipid under a gentle stream of nitrogen for at least 2 or 3 minutes. Make sure you can see the dried lipid.

3. To one of the vials, add 1 mL internal K solution. To the second vial, add 1 mL internal Ca solution containing Fluo-3 in the recommended concentration.
4. Sonicate the vials until cloudy.
5. Pipette the solution from each glass vial into a cryo vial. Freeze-thaw in liquid nitrogen three times. The thaw steps will take about 10 minutes each; you may proceed to the next step while you are waiting.

17.1.3.2 Creation of Ionic Imbalance and Potential Gradient

1. Pipette internal and external solutions into the black plate as shown in Figure 17.1.
2. After the last freeze-thaw step, pipette 30 μL of the lipid vesicles **without** Fluo-3 into each of the first two columns of the black plate. Pipette 30 μL of the lipid vesicles **with** Fluo-3 into each of the second two columns of the black plate. **Mix well.**
3. Add the voltage-sensitive dye (e.g., di-4-ANEPPS) to the first two columns as shown in Figure 17.1. Mix well.
4. Run a complete excitation and emission scan at the appropriate wavelengths.
5. Add valinomycin to the first column and calcium ionophore to the third column. Mix well by pipetting. Rerun the spectra immediately.

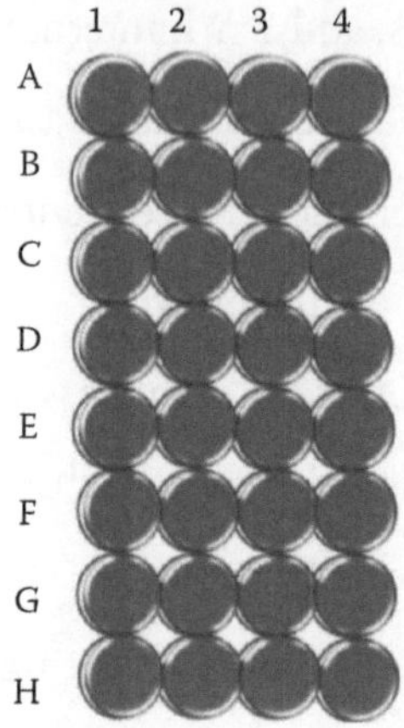

Figure 17.1 Final composition of plate, to be added in the order mentioned in the steps outlined in the text: **A1**: 270 μL K external, 30 μL vesicles without Fluo-3, ANEPPS, valinomycin; **B1**: 235 μL K external, 35 μL K internal, 30 μL vesicles without Fluo-3, ANEPPS, valinomycin; **C1**: 200 μL K external, 70 μL K internal, 30 μL vesicles without Fluo-3, ANEPPS, valinomycin; **D1**: 165 μL K external, 105 μL K internal, 30 μL vesicles without Fluo-3, ANEPPS, valinomycin; **E1**: 135 μL K external, 135 μL K internal, 30 μL vesicles without Fluo-3, ANEPPS, valinomycin; **F1**: 105 μL K external, 165 μL K internal, 30 μL vesicles without Fluo-3, ANEPPS, valinomycin; **G1**: 70 μL K external, 200 μL K internal, 30 μL vesicles without Fluo-3, ANEPPS, valinomycin; **H1**: 35 μL K external, 235 μL K internal, 30 μL vesicles without Fluo-3, ANEPPS, valinomycin; **A2-H2**: same as A1-H1 without valinomycin; **A3**: 270 μL Ca external, 30 μL vesicles with Fluo-3, calcium ionophore; **B3**: 235 μL Ca external, 35 μL Ca internal, 30 μL vesicles with Fluo-3, calcium ionophore; **C3**: 200 μL Ca external, 70 μL Ca internal, 30 μL vesicles with Fluo-3, calcium ionophore; **D3**: 165 μL Ca external, 105 μL Ca internal, 30 μL vesicles with Fluo-3, calcium ionophore; **E3**: 135 μL Ca external, 135 μL Ca internal, 30 μL vesicles with Fluo-3, calcium ionophore; **F3**: 105 μL Ca external, 165 μL Ca internal, 30 μL vesicles with Fluo-3, calcium ionophore; **G3**: 70 μL Ca external, 200 μL Ca internal, 30 μL vesicles with Fluo-3, calcium ionophore; **H3**: 35 μL Ca external, 235 μL Ca internal, 30 μL vesicles with Fluo-3, calcium ionophore; **A4-H4**: same as A3-H3 without calcium ionophore.

17.1.4 Expected Results and Pitfalls

The calcium-selective dye will provide a strong signal. Voltage-sensitive dyes show only subtle changes with voltage; observe at different excitation wavelengths to determine whether an effect can be measured. As mentioned in Chapter 16, it is important to use fresh ionophore solutions.

Background Reading

Nadeau, J. *Introduction to Experimental Biophysics: Biological Methods for Physical Scientists*. Boca Raton, FL: Taylor & Francis; 2012, Chapter 11.

Questions

1. Why must we freeze-thaw the vesicles?
2. Calculate the ionic gradients and Nernst potentials for all of the wells in the black plate.
3. What is the role of the ionophores?
4. Plot your fluorescence peak at a given wavelength versus Nernst potential. Comment.
5. Does the position of the peak change? If so, plot the position of the fluorescence peak versus the Nernst potential.
6. Do you expect the vesicle internal solution to be depleted? Why or why not? Over what time course?

APPENDICES
Instructor Information and Guides

Appendix 1 Disposables and Supplies Needed

A.1 In the –80°C Freezer

1. Aliquots of chemically competent cells
2. Aliquots of electrocompetent cells
3. Super optimal broth with catabolite repression (SOC) medium (2-mL aliquots)

A.2 In the –20°C Freezer

A.2.1 In Individual Freezer Boxes for Each Team

1. One aliquot of mystery plasmid DNA
2. Aliquot of ampicillin at 100 mg/mL, aliquot of kanamycin at 30 mg/mL
3. Restriction enzymes and buffers (small aliquot of enzyme for each team; one tube of each buffer)
4. T4 DNA ligase and buffer
5. Pellet paint
6. Electroporation cuvettes (3)
7. 1 mL of concentrated doxycycline stock

8. Polymerase chain reaction (PCR) primers
9. Deoxynucleotide triphosphates (dNTPs)
10. Polymerase

A.2.2 General Stocks to Share and for the Teaching Assistant

11. Midi-/maxiprep stocks of all of the plasmids used
12. Stocks of restriction enzymes and buffers
13. DOPC (dioleoylphosphatidylcholine) in chloroform
14. Dyes: di-4-ANEPPS, RH421, Fluo3

A.3 In the +4°C Refrigerator (Stocks to Share)

1. Luria-Bertani (LB) agar plates with kanamycin or ampicillin (4 per team) (just in case)
2. Lipofectamine (or other transfection reagent)
3. DNA ladder and protein ladder
4. Gel loading buffer
5. Valinomycin and calcium ionophore

A.4 On the Benchtop

A.4.1 For Each Team

1. One sleeve of Petri dishes
2. 500 mL Bacto Agar and 2X LB medium
3. Isopropanol
4. 70% ethanol
5. Agarose
6. Tris-acetate-EDTA (TAE) electrophoresis buffer

7. Gel tray, comb, and tape for agarose gel; casting plates for polyacrylamide gel electrophoresis (PAGE)
8. Razor blades
9. Ethidium bromide
10. 1.5 M Tris-HCl, pH 8.8
11. 0.5 M Tris-HCl, pH 6.8
12. 10% sodium dodecyl sulfate (SDS)
13. Sample buffer containing bromophenol blue and β-mercaptoethanol
14. Coomassie stain and destainer (methanol-40%, acetic acid-10%, distilled water-50%)
15. SDS-PAGE running buffer (Tris-glycine, pH 8.3)
16. Sterile water
17. 12-well mammalian cell culture plate
18. "Internal" and "external" lipid solutions
19. Glass vials, cryo vials
20. One black plate per 3 teams

A.4.2 Shared

Midiprep kit (1/team) and miniprep kit (at least 12/team) or reagents for home minipreps (as given in Experiment 2.1.3.2)

Spectrophotometer cuvette

Hamilton syringe

Decane (in the chemical cabinet)

Acrylamide solution (warning: neurotoxin)

Appendix 2 Instructors' Guide to Behind-the-Scenes Preparation

A2.1 Prior to the Course

Prior to the course, it is important to have midi- or maxipreps of all of the plasmids to be used. It is also helpful to have stocks of the successful ligation products, if these are available. This way, if everyone's ligation fails, there will still be something interesting to work with. At the very least, a pTre vector containing some sort of positive control reporter gene should be available.

It is highly recommended to make small aliquots of costly enzymes (restriction enzymes, polymerase, etc.) and to provide each team with only a small aliquot. Labeled freezer boxes with team numbers/names aid greatly in keeping reagents and minipreps separate. Large vials of restriction enzymes should not be passed around the class because they will be exposed to high temperatures, contaminated, or lost.

A good supply of ampicillin and kanamycin plates is also helpful. Sometimes, students fail to make any good plates, so they must be provided with plates.

A2.2 During the Course

Some of the experiments require behind-the-scenes intervention from the instructors to run smoothly. These are noted in the text and given here for easy reference.

- Experiment 1A: Remove plated cells from incubator after 12–16 hours and place in 4°C refrigerator.
- Experiment 1B: Remove cultures from incubator after about 8 hours and either pellet or store in the refrigerator.
- Experiment 3: Make sure all of the restriction enzymes the students request are available for the course. If not, suggest alternative strategies or buy the enzyme(s).
- Experiment 5A: It helps to provide computers during the downtimes (gel running, etc.) so students can work on cloning strategies.
- Experiment 5B: Remove cultures from incubator after about 8 hours and either pellet or store in the refrigerator.
- Experiment 6: It helps to provide computers during the downtimes (gel running, etc.) so students can work on predicting what they will see from the restriction digests.

- Experiment 7: Immediately after class, order polymerase chain reaction (PCR) primers from a selected vendor. Be aware of the vendor's lag time for scheduling Experiment 8.
- Experiment 8: There will be significant downtime as the PCR reactions run. This time might be used for lectures; alternatively, the students can come to purify their products the next day, in which case, the products should be removed from the PCR machine after the reaction and stored at 4°C. After purification, send the products with the appropriate primer for direct sequencing immediately. Use the known lag time of the chosen sequencing facility to schedule Experiment 8B.
- Experiment 9: It is highly recommended that transformation of BL-21 cells and induction be done by the teaching assistant. This saves an entire day of rather repetitive work for the students. If desired, transformation and plating of these BL-21 cells can be done as part of Experiment 1 along with the transformation of the mystery plasmids.
- Experiment 10: Precast gels may be used in the interest of time or to protect students from acrylamide exposure.
- Experiment 12A must be done on two successive days to allow the cells to attach after passaging. If this is not

possible, the passaging may be done by the teaching assistant (the students may be invited to watch outside class time if desired).

- Experiment 12B: Tet-inducible expression must be induced 1–12 hours before examining the cells; this can be done in class or by the teaching assistant before class.
- Experiment 13 must be done no less than 1 day and no more than 3 days after Experiment 12.

Appendix 3 Grading Recommendations and Guidelines

A3.1 Introduction

The following guidelines were distributed to the students before the course, accompanied by the university's general guidelines on academic honesty and plagiarism.

This is a graduate-level course and emphasis will be on (a) a laboratory notebook that would pass legal examination for intellectual property/authorship issues; and (b) writing part of a publication-quality article that uses the language and style of a biology journal.

A3.2 Grading

A3.2.1 Lab Notebook: 40%

Your lab notebooks will be collected at the end. They should contain an Introduction, Methods, and Results sections for each experiment. Write down everything you did, where things went wrong or right, and so on. In the case of an experimental failure, you will not lose points if you took good notes and can provide reasonable hypotheses for what went wrong.

A3.2.2 Homework: 30%

The homework questions are found in the lab notebook at the end of each chapter. You can write the answers in your lab notebook or on separate sheets. They are all due at the end of the course, but you may turn them in earlier as you wish. Do not get too far behind. These will be graded on a point system, like an exam, with 90% = A, 80% = B, and so on.

A3.2.3 Mock Paper or Final Exam: 30%

The "term paper" for this course will be a manuscript-style presentation of the Abstract, Introduction (with references), and Materials and Methods sections of a paper on one of the topics given in this appendix (or a topic of your choice; see the instructor first). There will be no Results or Discussion sections. You may use a template from any journal that you like; a useful one is found on the American Chemical Society website (http://pubs.acs.org/page/jacsat/submission/jacsat_templates.html) (follow Download Manuscript Templates). Due date is 1 week after the end of class. Late papers will lose one-third of a letter grade per week late.

Or, there will be a take-home final exam. All students must work individually, but any books, websites, or other references are permitted. Exams are due in 24 hours.

A.4 Sample Topics for Paper

1. A mutation in the chromophore region of green fluorescent protein (GFP) conferring novel fluorescence properties
2. Neuron-specific expression of GFP using the neuron-specific enolase promoter
3. Expression and purification of the *Escherichia coli* mechanosensitive channel and reconstitution into lipid vesicles
4. Properties of GFP fusions to the Shaker potassium channel

A.5 Sample Final Exam (Take Home)

Scoring: out of 100 (110 possible with bonus points)

A.5.1 Part 1: Molecular Biology

You can assume that you have access to all of the equipment and supplies of a typical molecular biology laboratory, as well as anything you might want to buy from catalogs.

Note: This is designed to be as real life as possible. All of these things have happened to your instructors, more or less. Each question may have more than one correct answer.

A.5.1.1 The Problem

You read a paper in *Science,* "New Cool Fluorescent Protein (CFP) Is Twice as Bright and Twice as Pretty as GFP." You want to try it, so you send an e-mail to the authors. You do not receive a reply to your message, but 2 days later you receive an envelope containing a piece of filter paper with a 1-cm circle on it, and the map shown below.

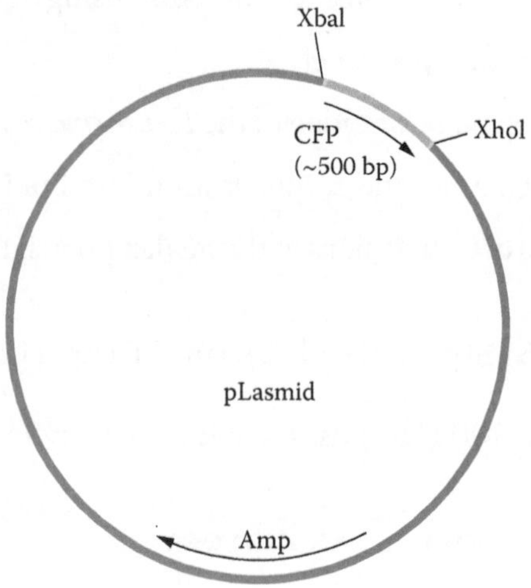

The filter paper kind of freaks you out, but you remember that you only need nanograms of a plasmid to obtain colonies. You soak the circled part in an Eppendorf tube with some water for half an hour, transform it into chemically competent cells, and plate on ampicillin plates. Sure enough, the next day you have colonies.

1. What is the first thing you should do? (5 points)
2. What is the first screening test you should perform? (5 points)

3. You transfect some of the DNA into some 293 cells you have around, but you do not see any fluorescence. Give at least two explanations for why this could be true. (10 points)
4. You cannot find plasmid in any catalogs, and when you e-mail the authors, they do not answer, so you decide to cut the gene out and put it into your favorite vector. Interpret the possible meanings of the gels shown in the following figure. (15 points)

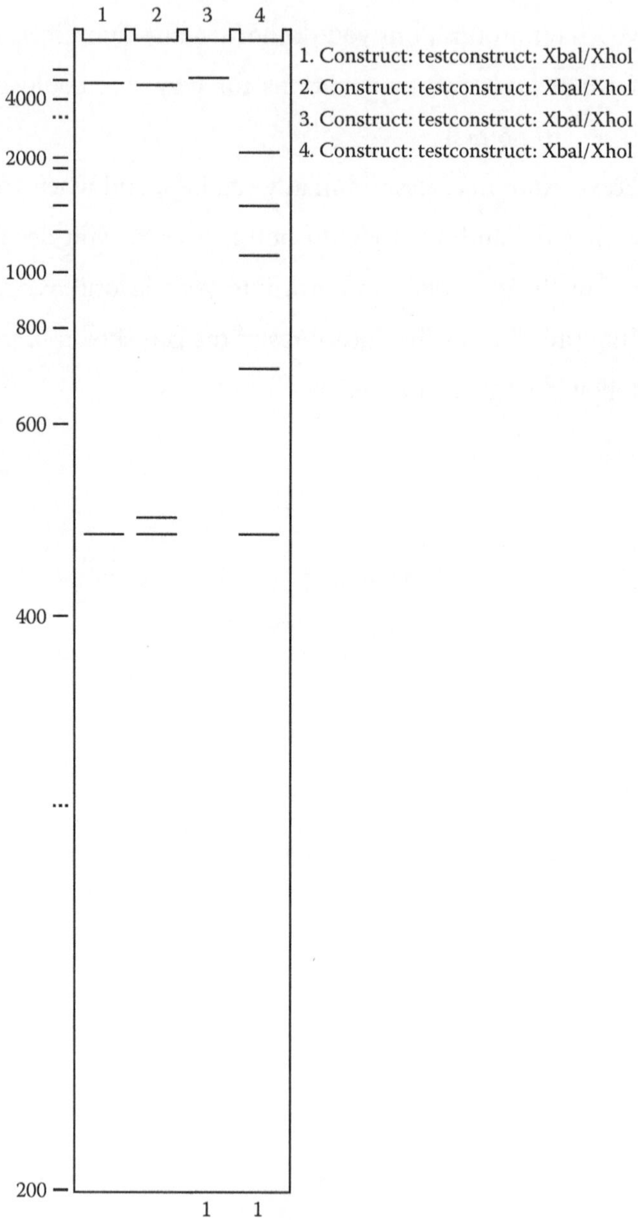
1 2 3 4
4000
2000
1000
800
600
400
200
1 1
1. Construct: testconstruct: XbaI/XhoI
2. Construct: testconstruct: XbaI/XhoI
3. Construct: testconstruct: XbaI/XhoI
4. Construct: testconstruct: XbaI/XhoI

5. How will each of these results affect your cloning strategy? (15 points)
6. At what point would you just give up and use a polymerase chain reaction (PCR) to figure it out? How would you design primers? (5-point bonus)

PART 2: PROTEINS AND ELECTROPHYSIOLOGY

The Problem

You turn the page of your *Science* magazine, and you see "New Great Big Cation Channel (GBCC) is open for 5 s and has a conductance of 250 pS." The article tells you how to purify the His-tagged protein and how it behaves electrophysiologically (open only at negative voltages, blocked by pufferfish toxin TTX). You really want some, so you send an e-mail to the authors; this time, you are not freaked out by the filter paper you receive in the mail. The authors send you a better plasmid map than the CFP group shown in the figure.

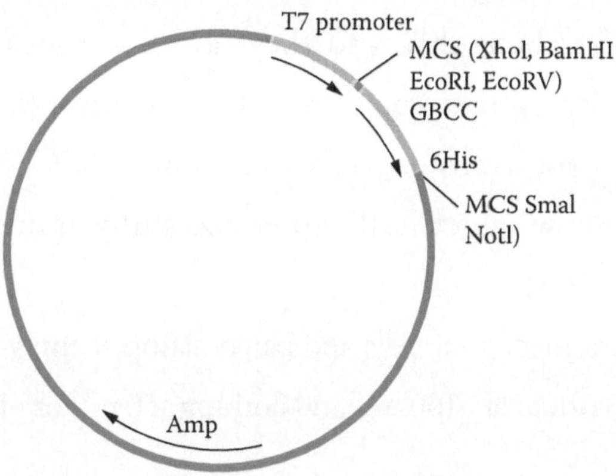

1. What is the first thing you should do? (5 points)
2. What is the difference between amplifying *plasmid* and amplifying *protein*? (5 points)
3. Give an outline of the shopping list that you would need to grow and purify the His-tagged protein (bacterial strain, columns, etc.). (5-point bonus; not discussed in class)
4. By a miracle, you think your purification worked. You set up your bilayer chamber and put some in. What would you expect to see in 1 mM KCl at –70, 0, and +70 mV? Draw a trace with a time and current scale bar (this requires you to convert conductance in pS into current in pA—a classic ion channel problem). (10 points)
5. How would you really convince yourself that this is the GBCC and not just a leak? (Hint: There are several correct answers.) (5 points)
6. You get sick of the bilayer chamber and want to try expressing it in mammalian cells. You ask the authors about this, and they say, "Cool idea, never tried it." What would you do to make this work? When you transfect cells with your GBCC, how do you know which cells are successfully transfected? (10 points)
7. You transfect your cells and patch-clamp them; you see a huge current at –100 mV and nothing at 0 mV or +100 mV.

What is this huge current? How can you convince yourself that it is because of the GBCC? (5 points)

8. Answer one of the following questions (5 points):
   a. How would you study single-channel properties in your mammalian cells? Please do not just name a technique; provide a few words about why you would use that technique.
   b. How would you make a fusion of GBCC to GFP, and how would you test to see if it works? What kind of effects would you expect GFP might have on the channel?

Appendix 4 Advanced Options: Customizing the Course

An advanced option may be added to the course in which students are encouraged to bring their own genes (usually graduate students using genes from their research projects). These genes may be used in the series of experiments described here: put under tetracycline (tet) control, fused to green fluorescent protein (GFP), and so on. Depending on the identity of the gene, it may be incorporated into the more advanced experiments (e.g., used for electrophysiology).

Example A4.1

A student is working with the Shaker potassium channel. A cloning strategy is designed to fuse enhanced green fluorescent protein (EGFP) to the channel's terminus. The channel is then used in a mammalian cell electrophysiology demo or class experiment.

Example A4.2

A student is working with a gene that kills cells when transfected. The student places this gene under the control of the tet promoter and cotransfects with a fluorescent protein. The student then titrates with doxycycline to determine levels of expression that will not kill cells.

Example A4.3

A student is working on a crystallization project. The gene is expressed in BL-21 cells and the protein purified. A strategy for crystallization is developed, and a first iteration of parameters may be performed, with scoring of the resulting crystals.

Index